ST. MARY'S CITY, MARYLAND 20686

STARTING WITH SYNFUELS

A REPORT OF HARVARD'S ENERGY AND
ENVIRONMENTAL POLICY CENTER

STARTING WITH SYNFUELS
Benefits, Costs, and Program Design Assessments

JAMES K. HARLAN

BALLINGER PUBLISHING COMPANY
Cambridge, Massachusetts
A Subsidiary of Harper & Row, Publishers, Inc.

Major parts of this book are based on a study which the author prepared during 1980 for the United States Department of Energy, in his capacity as an employee of the United States Government. As a work of the United States Government, the study from which parts of this book are derived is not subject to copyright.

Any opinions, findings, conclusions, or recommendations contained in this book are those of the author and do not necessarily represent the views of the United States Department of Energy or any other agency of the United States Government.

Copyright © 1982 by Ballinger Publishing Company. All rights reserved. No part of this publication may be reproduced, stored in a retrieval system, or transmitted in any form or by any means, electronic, mechanical, photocopy, recording or otherwise, without the prior written consent of the publisher.

International Standard Book Number: 0-88410-869-4

Library of Congress Catalog Card Number: 81-22824

Printed in the United States of America

Library of Congress Cataloging in Publication Data

Harlan, James K.
  Starting with synfuels.

  Includes bibliographical references and index.
  1. Synthetic fuels industry—Government policy—United States. I. Title.
HD9502.5.S963U544   338.4'766266'0973   81-22824
ISBN 0-88410-869-4   AACR2

CONTENTS

List of Figures — xi
List of Tables — xv
Foreword — xix
Preface — xxiii
Acknowledgments — xxxi

Chapter 1
Introduction and Overview — 1

The Paradox of Certain Potential and Uncertain Role — 1
  Certainties and Uncertainties — 1
  Surges of Interest — 2
Assessing the Social Investment in Synthetic Fuels — 3
  Three Basic Questions — 3
A Government Program and the Free Market — 9
Emphasis and Limitations of This Study — 10
  Generic, Coal-Based Synthetic Fuels — 10
  Two Issues Not Addressed — 10
  How This Study Differs From Other Synfuels Studies — 11
Assessing the Synfuels Insurance Investment — 13
Notes — 14

PART I. WHETHER AND WHY A SYNFUELS PROGRAM? 15

Chapter 2
Benefits from Initial Synfuels Production: A Framework and Assessment 17

Introduction 17
Production Benefits 18
   Benefits From an Oil Import Premium 19
   Energy Security and Synthetic Fuels 22
Information Benefits from Initial Synthetic Fuels
Production Experience 28
   Improved Information for Better Decisions 28
   Potential Oil Pricing Impacts of Information from
   Synthetic Fuel Production 31
Surge Deployment Benefits 34
   A Key Source of Strategic Insurance Benefits 34
   Learning Benefits for Surge Deployment 38
   Enhanced Deployment Capability 47
   Improving the Growth Basis for Surge Deployment 64
Information and Capabilities as Strategic Insurance 66
Notes 68

Chapter 3
Costs for Initial Synfuel Production Efforts: Basic Elements and Assessments 73

Introduction and Overview 73
Net Production Subsidy Costs 76
   Ranges for Synthetic Fuel Production Costs 77
   Oil Price Trajectories and Petroleum Product Costs 89
   Estimates of Net Production Subsidy Costs 92
Factor Cost Inflation 98
   Sources of Factor Cost Inflation: A Qualitative Overview 98
   Estimating Factor Cost Inflation by Analogy 100
"Whether?" A Synfuels Program: Conclusions About Costs and
Comparisons with Benefits 104
   Two Qualitative Conclusions 104
   Example Estimates of Synfuel Program Costs 105
   The Paradox of Synfuel Costs, Savings, and Payoffs 106
   The Key Role of a Framework for Benefits and Costs 107
Notes 108

PART II. "HOW LARGE?" A SYNFUELS PROGRAM: THE DECISION ANALYSIS ASSESSMENT 113

Chapter 4
General Structure of Synfuels Decision Analysis Framework and Model 119

Structuring a Decision Analysis Framework For A Simple Insurance Problem 119
General Structure of the Synfuels Decision Analysis 123
   Decade-by-Decade Decision Structure 124
   Key Uncertain Outcomes 126
Modeling for Particular Elements of the Decision Tree 128
   Program Scale Decision in 1980 (*PSD*–80) and Factor Cost Inflation 129
   Long-Run Oil Price Trends (*OP*–80, *OP*–90) 131
   Basic Synfuel Production Cost Outcome (*SC*) 134
   Program Benefit Outcome Nodes (*PBO-L, PBO-D*) 135
   Surge Deployment Events (*SDE-P, SDE-S*) 140
   1990s Production Expansion Decision (*PED*–90) 144
   Synfuel Production Limit (SPL) 145
   Other Outcomes 146
   Overall Decision Tree 149
Calculation of Benefits for Alternative Scenarios 149
   Summary 151
Notes 151

Chapter 5
How the Model Works 155

Calculating Benefits and Costs for a Particular Scenario 155
Weighting Scenario Estimates by Probabilities 157
Comparing Appropriate Program Size for Two Extreme Scenarios 158
Relating These Examples to the "Full" Model Results 160
Notes 162

Chapter 6
The Decision Analysis Results 163

Using the Decision Analysis Tool 163
Base Case Results 164

Questions Addressed by Model Outputs	164
Basic Model Results	165
A Qualitative Interpretation of Results	167
Results for Particular Synfuel Cost and Oil Price Outcomes	168
How Much Is Better Information Worth?	170
Preferences for Particular Views of Oil Prices and Synfuel Costs	174
Results of Sensitivity Analysis	175
The Key Role of Sensitivity Analysis	175
Overview of Groups of Sensitivity Experiments	176
Varying Probabilities for Surge Deployment Events	179
Exploring the Importance of Various Sources of Benefits	184
How Factor Cost Inflation Affects Recommended Choices	191
Alternative Descriptions for Surge Deployment Events	203
Sensitivity Cases for the Value of Infrastructure Development	207
Are Free Market Incentives Alone Adequate?	211
Using Sensitivity Cases	217
Notes	218

Chapter 7
Summary Observations on "How Large?" A Synfuels Program — 225

Bigger Is Not Necessarily Better	225
Preferred Program Size Is Sensitive to a Few Key Assumptions and Effects	226
Synfuel Costs Relative to Long-Run Oil Prices	227
Likelihood of Surge Deployment	227
Factor Cost Inflation	227
Value of Infrastructure Development	228
Moderate Program Is Robust Choice	229
Conclusions for the "How Large?" Question	233

PART III. HOW TO DO BETTER? — 235

Chapter 8
Program Design Issues for Enhancing Learning — 239

Introduction: Two Types of Learning	239
Selection Learning: Design Issues and Analysis	240

Sound Differentiation and Choice	241
Significant Future Options	246
Available and Interpretable Information	249
Analysis Steps for Improving Selection Learning	251
Technology Improvement Learning: Descriptions and Design Issues	254
The Process for Technology Improvement Learning	255
Issues for Enhancing Technology Improvement Learning	264
"Doing Better" for Learning: Four Lessons	269
A Properly Phased Program Greatly Enhances Learning	270
Diversity Fosters Learning	270
Information Is a Key to Learning	271
Smaller Programs and Projects May Foster Learning	271
The Importance of Defining Priorities Among Objectives	272
Notes	272

Chapter 9
Building Deployment Capability: Strategic Issues and Analytical Approaches — 275

A Strategic Approach for Assessing Deployability Issues	277
Separating Long from Short Lead-Time Concerns	278
Understanding How Production Experience Relaxes Constraints	279
Service Inputs	281
AE&C Services	281
Construction Labor	286
Hardware Inputs: Equipment, Material, and Components	288
Defining Potentially Constrained Items	289
Assessing Strategic Investments To Relax Hardware Constraints	289
Most Hardware Constraints Are Short Lead-Time Issues	291
"Doing Better" for Infrastructure Development	292
Long Lead-Time Issues Provide Focus for Program Design	292
AE&C Capabilities Are a Key Long Lead-Time Input	293
Payoffs from Infrastructure Development Depend on Expectations for Our Energy Future	293
Issues for "Doing Better"	294
Notes	294

Chapter 10
Applying This Analysis to the U.S. Synthetic Fuels Corporation Program — 297

The Energy Security Act Program — 297
   Basic Elements of the Program — 297
   Tensions Among Multiple Purposes — 298
Relating This Analysis to the SFC Program — 299
   Program Purposes — 300
   Phased Production Goals and Timing — 302
   Comprehensive Production Strategy — 304
   Program Design and Implementation Provisions for "Doing Better" — 305
   SFC Organizational Capabilities — 307
Conclusions for the Energy Security Act Program — 308
Notes — 310

Chapter 11
Conclusions — 313

A Rationale and Strategy for Getting Started — 313
National Purposes Need Not Imply Public Management — 315
Lessons for a Strategy for Getting Started — 318
   Go, But Go Moderately — 319
   Factor Cost Inflation Is a Critical Issue — 320
   Infrastructure Benefits Tradeoff with Factor Cost Inflation — 321
   Use a Truly Phased Program Structure — 321
   Collect Information — 322
   Respond to Information — 323
   Consider the Advantages of Smaller Projects — 324
   Emphasize Future Capabilities with the Synfuels Option, Not Near-Term Production Levels — 324
Notes — 324

References — 325

Index — 331

About the Author — 335

LIST OF FIGURES

2-1	Stylized Hypothetical Learning Curve	42
2-2(a).	Learning Curve/Capacity Relationship for Surge Deployment Effort	43
2-2(b)	Surge Deployment Cost Levels Reduced via Learning	43
2-3	Illustration of Concept for Enhanced Deployability Benefits	50
2-4	Illustration of Stylized Modeling for Price Jump Event	58
3-1	Increasing Marginal Costs with Increasing Program Size Due to Factor Cost Inflation	105
4-1	Decision Analysis Structuring of Simple Flood Insurance Problem	121
4-2	Major Elements of the Synfuels Insurance Problem	124
4-3	Three-Decade Structuring for Synfuels Decision Analysis	125
4-4	Representation of Uncertain Outcome Node and Branches	126
4-5	Structure of Decision Tree for 1980s Decade	128
4-6	1980 Department of Energy Policy and Fiscal Guidance Long-Run Planning Oil Price Trajectories	133
4-7(a)	Contrasting Relationship of Program Benefits to Program Size: Learning and Information Benefits	136
4-7(b)	Contrasting Relationship of Program Benefits to Program Size: Deployability Benefits	136

4-8	Full Decision Tree Structure	148
4-9	Illustration of Cost-Quantity Benefit Calculation and Sources of Benefits	150
5-1(a)	Results for Two Contrasting Scenarios	159
5-1(b)	Probability-Weighted Result for Two Extreme Scenarios Each Having Equal (50 Percent) Probability	161
5-1(c)	Probability-Weighted Result When High Synfuel Benefits Scenario Is Twice as Likely to Occur	161
6-1	Standard Base Case Results	166
6-2	Qualitative Interpretation of Standard Base Case Results	169
6-3	Base Case Results for Particular Oil Price Trends and Synfuel Cost Outcomes	171
6-4	Sensitivity Cases Varying Likelihood of Security-Motivated Surge Deployment Event (*SDE-S*)	181
6-5	Sensitivity Cases Varying Likelihood of Oil Price Jump Events (*SDE-Ps*)	184
6-6	Sensitivity Cases Separating Learning and Deployability Effects	187
6-7	Varying Specifications for Base Deployment Capability	189
6-8	Sensitivity Cases for Alternative Descriptions of Program Benefit Relationships	190
6-9	Sensitivity Cases Varying Amounts of Factor Cost Inflation	194
6-10	Similar Shape Net Benefits Curves for Alternative Modeling of Factor Cost Inflation	196
6-11	Illustration of Crossover Amounts of Factor Cost Inflation that Equalize Benefits of Different-Size Programs	198
6-12	Variation in Acceptable Levels of Factor Cost Inflation (FCI) with Likelihood for Surge Deployment Events	201
6-13	Sensitivity Cases Varying Amount of Oil Price Jump	206
6-14	Sensitivity Cases Varying Length of Price Jump Event (Length of Decline Period Back to Long-run Trend "LDP")	206
6-15	Sensitivity Cases for Modeling of a Security Event (*SDE-S*)	207
6-16	Sensitivity Cases Illustrating Effect of Synfuel Production Limit Outcome on Value of Larger Programs	210

8-1(a)	Hypothetical Cost Ranges/Distributions Before Experience	243
8-1(b)	Hypothetical Cost Ranges/Distributions After Initial Production Experience	243
8-2	Illustration of Relative Amounts and Locations of Learning During Progress of a Synfuels Project	259

LIST OF TABLES

2-1	Example Benefits from Selection Learning	41
2-2	Potential Per Barrel Amounts of Technology Improvement Cost Savings	45
2-3	Benefits Per Plant from Technology Improvement Learning	45
2-4	Example Synfuel Production Levels with Enhanced Deployability	54
2-5	Department of Energy Policy and Fiscal Guidance Price Trajectories	57
2-6	Net Present Value for Each Surge Deployment Plant	59
2-7	Net Present Value for Each Surge Deployment Plant	61
3-1	Representative Estimates of Plant Capital Investment Costs for a Coal-Based Synfuel Plant	82
3-2	Factors for Estimating Fixed Operating and Maintenance Costs	83
3-3	Net Feedstock Costs for Varying Coal Costs and Coal-to-Products Conversion Efficiency	84
3-4	Effect of Varying Plant Capacity Factor on Per Unit Synthetic Fuel Costs	87
3-5	Ranges for Components of Total Coal-Based Synfuel Production Costs	88
3-6	Constant Percentage Annual Growth Rate Oil Price Cases	90

LIST OF TABLES

3-7	Price Schedules for Example Sensitivity Oil Price Trajectories	91
3-8	Form Value Differentials for Petroleum Products	93
3-9	Production Subsidy Estimates for Department of Energy Policy and Fiscal Guidance Oil Price Trajectories	95
3-10	Production Subsidy Estimates for Annual Percentage Growth Oil Price Paths	96
3-11	Production Subsidy Estimates for Sensitivity Case Oil Price Trajectories	96
3-12	Subsidy with Varying Discount Rates for Department of Energy Policy and Fiscal Guidance Price Trajectories	97
3-13	Subsidy for Varying Plant Initial Production Date	98
3-14	Example Calculations for Increasing Synthetic Fuel Production Costs if Hyperinflation Similar to 1973-74 Period Is Experienced	102
4-1	Base Case Specifications for Program Size and Factor Cost Inflation	131
4-2	Base Case Probabilities for Oil Price Trends	134
4-3	Base Case Assumptions and Probabilities for Synfuel Cost Outcomes	135
4-4	Base Case Probabilities and Values for Learning Outcomes	138
4-5	Base Case Expectations for Magnitude and Probability of Deployability Increase Outcomes	141
4-6	Probabilities for Surge Deployment Events in Each Decade	144
4-7	Base Case Specificiations for Synfuel Production Limit Outcomes	147
6-1	Expected Benefits of Initial Synfuel Production Programs for Base Case	165
6-2	Relative Value of Alternative Initial Production Programs for Particular Synfuel Cost Outcomes	170
6-3	Relative Value of Alternative Initial Production Programs for Particular Long-Run Oil Price Outcomes	170
6-4	Relative Program Benefits for Various Combinations of Long-Run Oil Price and Synfuel Cost Outcomes	174
6-5	Sensitivity Cases Varying Probability for Security-Motivated Surge Deployment Events (*SDE-S*)	180
6-6	Sensitivity Cases Varying the Probability for Price Jump Events	182

6-7	Probability Ranges for Price Jump Events for Which Particular Program Sizes are Recommended	185
6-8	Sensitivity Cases Illustrating Relationships Among Sources of Program Benefits	186
6-9	Sensitivity Cases for Factor Cost Inflation	193
6-10	Amounts of Benefit Reduction Due to Factor Cost Inflation	197
6-11	Cross-Over Levels of Factor Cost Inflation	199
6-12	Sensitivity Cases for Alternative Descriptions of Price Jump Events	204
6-13	Sensitivity Cases Illustrating the Value of Infrastructure Development	209
6-14	Sensitivity Cases Illustrating Potential Divergence Between Nation's Choice and Private Incentives	214
7-1	Summary of Sensitivity Case Results for "Benefits Foregone" Percentages	231
8-1	Time Required to Implement a Commercial Synfuels Project	257
8-2	Attributes of Learning Process for Synthetic Fuels	263
9-1	Materials and Equipment Items Potentially Constraining a Major Synfuels Deployment Effort	290

FOREWORD

When President Carter decided halfway through his term to deregulate domestic crude oil, phasing controls out over two and a half years, the prospective rise in domestic oil revenues made some tax inevitable. What Congress enacted, usually called the "windfall profits tax," was an excise on the increase in the price of crude oil. Revenues for the decade were estimated in the range of from $100 billion to $200 billion at the OPEC prices of that time, with every likelihood that prices and revenues would be even higher. (The huge increases that were actually to occur by the end of 1980 might have trebled the estimates had they been foreseen.) The question naturally arose, what to do with the proceeds?

I say "naturally," not "logically." By any responsible principle of public finance, such huge revenues should have been treated as real money, not as a windfall to be splurged like some winnings at bingo. But casuistry, politics, and perhaps a measure of opportunistic sound policy combined to identify these funds that originated in energy with programs directed toward energy.

There was no possibility of spending that kind of money on research, subsidies for home insulation and solar collectors, or fuel assistance to the poor. Some major undertaking was called for, one with a face value approaching $100 billion and preferably one directly related to American dependence on oil from abroad. The only two candidates—a natural gas pipeline across Canada not yet having achieved eligibility—were nuclear electricity and synthetic liquid fuels. And at that historic moment nuclear power didn't have a chance.

Synthetic liquids offered several advantages. Liquid fuel can do almost anything that gas and coal can do, but not vice-versa. Transportation especially has been largely confined to liquid fuels for many decades. Nuclear power, even had it not at that time been troubled by safety and waste disposal, can produce only electricity, and EPA permitting, even coal can do that. Liquid fuels are what make us and our allies susceptible to the politics of the Middle East and militarily vulnerable in the Persian Gulf. Finally, not enough was known about the likely costs, environmental impacts, or schedules for completion of full-scale facilities to produce liquid fuels from coal or shale. Hence, it was impossible to put realistic limits on the capital outlays and the target production rates that might plausibly go into a program aimed at the early 1990's.

"Face value" was important in setting aside these oil revenues. It was never intended that the government invest directly in synthetic fuels plants and become owners and operators. The intention all along was to offer loans, or to guarantee loans, or to provide price assurances, or to invent other forms of financial backing that would induce private investment and private management to fulfill the program's goals. At most the net cost to the U.S. government would be the cost of abandoned plants, the excess cost of plants that upon completion couldn't compete at market prices, some direct expenditures for development, or in rare cases some sharing in equity. But the program served to charge a considerable capital outlay as a contingent claim on a large fraction of the oil revenues that were then foreseen, providing at least a transient answer to the question of what to do with the oil revenues.

It had never been at all clear what would constitute a proper strategy toward the development of a *productive capacity* for synthetic fuels. Theories of research and development could encompass the construction of pilot plants and small-scale facilities; and there is no great embarrassment about parallel development of alternative technologies, some of which will be abandoned as they lose in competition with other technologies. But proceeding simultaneously with "pioneer" facilities on a scale large enough to simulate actual production—while developing the industrial base for rapid expansion of capacity and bringing actual fuel to market within a decade on a scale equivalent to Alaskan production—was bound to put awesome demands on the strategic planners, for whom Congress was setting aside $20 billion as a first installment.

It is gratifying to anyone concerned with policy research to see a project like this reach completion just as the demand for it ripens into urgency. Before Congress enacted the Energy Security Act of 1980 and even before the president launched his proposal, Jim Harlan perceived that some active policy toward synthetic liquids was a likely possibility. Marginal subsidies, tax benefits, and price guarantees were continually being proposed and discussed. He saw what the issues were, especially the fundamental ones. He brought to the subject relevant technical training and experience, some of it earned while working on energy research and development in the Department of Energy itself, some through the training at the John F. Kennedy School of Government, Harvard University.

Here is the book that Jim Harlan completed just as the directors of the new Synthetic Fuels Corporation began to ask urgently the questions that this book answers.

Thomas C. Schelling
Lucius N. Littauer Professor
  of Political Economy
John F. Kennedy School of
  Government
Harvard University
Cambridge, Massachusetts

PREFACE

The Energy Security Act of 1980 authorized the expenditure of up to $88 billion dollars to promote initial synfuel development. This book addresses basic questions concerning such a program: Should the nation make special investments to promote initial synfuel production projects? Are market incentives alone adequate? How large should any program be? What are cost-effective objectives? A balanced assessment of these important questions requires integrating the perspectives of economists, engineers, strategic planners, and investment analysts. Also, the many paradoxes and contrasts surrounding synthetic fuels should be recognized.

Synfuels pose contrasts of certainty and uncertainty. Technology and resources certainly exist to produce enormous quantities of synfuels. However, the costs of realizing the undeniable potential for synfuels production remain uncertain and the technologies untried. Technologists, politicians, and some energy planners have focused on the certain potential of synfuels to advocate building a large industry to reduce oil imports. On the other hand, economists, industrialists, private investors, and other energy analysts, observing the uncertainties and costs, have been less enthusiastic about a large near-term synfuels industry or even an initial production program. The spectrum of views regarding a synfuels industry, as well as what role government should assume in promoting initial projects, ranges widely.

For many Americans the notion of combining domestic resources with technology to displace unreliable imported energy has understandable appeal. However, the likely high costs of coal-based synfuels pose another paradox that strikes the pocketbooks and balance sheets of energy consumers. If synfuels cost much more than oil (say $60/bbl or more), should we really hope that a synfuels industry becomes generally economic soon, if ever?

The certainties, uncertainties, costs, and various contrasting views regarding synfuels have contributed to a "boom and bust" cycle of enthusiasm for government programs to promote initial production projects. The oil embargo and price increases of 1973-74 fostered "Project Independence" and serious consideration of legislative proposals authorizing a major synfuels program. However, uncertain costs, environmental impacts, and divergent views regarding the federal role for synfuels development, combined to discourage Congress from enacting those proposals during the subsequent period of oil price and supply stability between 1975 and 1979. A second "boom" period for synfuels development followed the major oil price increases of 1979 and heightened awareness of the political instability in the Persian Gulf. The fervor surrounding the events of 1979-80 fostered enactment of the Energy Security Act of 1980, which established ambitious synfuel production goals backed by appropriations of $20 billion for synfuels development incentives and authorizations for an additional $68 billion. However, as we move into the 1980s, the predictable marketplace "glut" as supply and demand respond to a major price increase, budgetary pressures, and a new administration skeptical of the merits a major federal role in energy markets, may be combining to suggest the possibility of another "bust" cycle for a major synfuels program. At least, some reassessments are likely.

These surges of enthusiasm and reassessment regarding a government synfuels program have been separated by less time than it takes to plan, design, and construct a typical synfuels project. Synfuels projects require enormous capital investments that payoff over a long period of time. The private sector routinely makes such long-term commitments for clearly economic investments. However, if synfuels are not yet economic in the marketplace (and one might hope they do not soon become economic because of oil price increases), government commitments to economic incentive programs must also be stable if they are to be credible inducements to private investors. The cyclical history of government commitment to a synfuels program poses another uncertainty for private project developers. There certainly is a government program now. But will the commitment outlast the time frame for project development and payoff?

This book seeks to address some of these contrasts and questions in order to articulate and analyze a stable economic and strategic basis for a prudent national program to promote initial synthetic fuels production projects. The book is motivated by the concern that the long-term investment commitments required by both the private and public sector for a synfuels program are better based on clearly articulated and assessed economic and strategic interests than on the fervor of a passing crisis, or even the commitment of a changeable political leadership.

Answers to the basic questions addressed by this book are not clearcut. Many arguments of both synfuels advocates and critics have merit. A balanced but questioning view is needed. For example, synfuels advocates frequently offer the energy security costs of oil imports as rationale for a program, without explaining how synfuels production will enhance energy security or asking whether an equivalent amount of domestic oil production, conservation, or strategic storage would provide a similar improvement in energy security more cost-effectively. On the other hand, some economists and industrialists argue the adequacy of free market incentives without assessing the public benefits from pioneering synfuel projects, the strategic dynamics which are all too possible in oil markets, or the technical practicalities of building capability for a synfuels industry. This study seeks to articulate and assess an underdeveloped middle ground which integrates contrasting views. The concepts developed in this book are not really new. The book primarily articulates, quantifies, and compares sources of benefits and costs that have long been recognized.

Quantitative analysis of costs and benefits is an important aspect of this book. The question of whether to have a federal synfuels program and how large it should be is an important issue for the allocation of our limited resources. Just as corporate stockholders and boards expect careful assessments of the expected return on private investments, the investment of billions of taxpayer dollars to foster initial synfuel projects should be based on an explicit consideration of expected benefits, costs and returns. Some of the benefits and costs which should be factored into government's broad national view are difficult to quantify. However, this study shows that many important benefits and costs can be quantitatively addressed. Moreover, the book seeks to illustrate how some of the less quantifiable issues can be considered by asking how a given objective can be most cost-effectively realized.

The quantitative, analytical orientation of this book may well be tedious to some readers, illuminating to others, and obvious to still others. The qualitative strategic concepts and framework can be grasped without numbers. However, the many contrasting costs, benefits and scenarios can quickly undermine the sureness of one's intuition. Systematic quantitative analysis can help confirm or test intuitive assessments. Such probing can contribute to stable, well-grounded assessments and decisions. Thus, the presentation maintains a quantitative orientation with a conscious effort to allow the numbers and analysis to be readily interpretable to interested, but non-analyst readers. Time and words are taken to enhance the interpretability (and, one hopes, the credibility) of the analysis by explaining important assumptions, calculations, and sensitivities.

This book may interest diverse audiences ranging from the broadly concerned citizen, to synfuel project developers, to public policy officials and analysts, to operations research specialists. Thus, many readers may be more interested in the qualitative conclusions and implications of quantitative analysis than in the procedure and details underlying those conclusions. Other readers may be more concerned with analytical details including the practical application of cost-benefit analysis and decision analysis to a major public investment decision. Accordingly, the presentation is divided into three parts. Each part addresses separate questions, has a different level of quantitative emphasis, and may hold varying interest for different audiences.

Part I addresses the question of whether and why a synfuels program may be warranted. It considers major sources of benefits and costs and develops illustrative estimates of how those benefits or costs may change with various scenarios or assumptions about key uncertainties. The qualitative strategic framework for sources of benefits and costs is critical to understanding the analysis and argument in this book. The illustrative quantitative estimates show that potential benefits could be large, small, or about the same size relative to costs. The book endeavors to make the basis for such estimates clear, but readers are encouraged to skim or skip discussions of the calculations, depending on their interests.

Part II considers the appropriate size of a synfuels program. Assessments of this quantitative question naturally emphasize quantitative analysis. Accordingly, much of the presentation in Part II is devoted to describing the structure, operation, and results for an analytical tool (a "model") which helps calculate, summarize, and compare key benefits and costs for the many assumptions and scenarios which bear on the "how large?" question. Here again, the presentation seeks to make the logic and calculations underlying the quantitative results readily interpretable to interested readers who are not experts with the analytical skills required. This effort adds length for some technically-oriented readers, but the explanations add clarity as well. Some of Part II's length also stems from the large number of sensitivity cases which are explored in order to provide some quantitative feel for how preferences on the "how large?" question change with varying assumptions about key uncertainties. Much of Part II can be skimmed or skipped by readers not interested in understanding the details underlying the quantitative analysis. Key qualitative conclusions from the analysis are summarized briefly in Chapter 7.

Part III shifts away from a quantitative emphasis to address the largely qualitative question of how to do better with whatever initial synfuel production program is undertaken. Many readers less oriented toward quantitative analysis may be primarily interested in Part III's review of the approaches and issues for cost-effectively realizing key program benefits and purposes.

As with much of this book, Part III explores program design and implementation issues for a general initial production program without detailed emphasis on the particular program authorized by the Energy Security Act of 1980. This generic and conceptual emphasis is purposeful. It reflects the view that normative design of a long-term strategic investment—such as a synfuels program—should be based on a fundamental understanding of ways to realize important program objectives, rather than primarily the details of a legislative debate or a particular administrative approach to implementation of the resulting law.

Chapter 10 departs from this generic emphasis to consider the Energy Security Act program in particular. The structure, production goals, and timing of the ESA program are reviewed in light of the lessons of the quantitative and qualitative analysis. More importantly, Chapter 10 reviews some of the implications of this analysis for the wide range of program implementation choices available to the Synthetic Fuels Corporation (SFC). Readers primarily interested in the "bottom line" implications of this assessment for the SFC program should focus on Chapters 10 and 11.

Teachers and students of public policy analysis may observe that the book's three part structure follows an established paradigm for public policy analysis. Part I's review of benefits, costs, and possible market failures seeks to define critically a well-grounded rationale for government action. Part II's quantitative assessments of benefits and costs consider whether investment in pioneering synfuels projects is likely to be worthwhile for the nation. It also asks whether private investments in response to market opportunities alone are likely to serve the national interest adequately. Extensive sensitivity analysis is presented because appropriate public investment choices may change with alternative views of key parameters, probabilities, and assumptions. Part III's review of implementation issues seeks to identify means of making any public investment program to promote initial synfuel projects more cost-effective.

A note on the genesis of this book may clarify its somewhat conceptual treatment of an ongoing, statutorily prescribed public investment program. Background research for this book was begun in late 1979 and continued through 1980 as a special project for the U.S. Department of Energy. DOE had a practical interest in assessing the benefits and costs of a commercial synfuels program, as well as measures for making any program more cost-effective. Quantitative assessment of some of the programmatic questions fit neatly with the interests and skills of a group at the Energy and Environmental Policy Center of Harvard's John F. Kennedy School of Government. The academic environment encouraged focus on the basic concepts and public policy questions—Whether? Why? How large? How to be cost-effective?—even as the legislative process for the Energy Security Act was providing statuatory guidance on some of these questions. Thus, the work sought to strike a balance between long-term fundamental concepts and ongoing legislative and administrative developments. During 1981, Harvard's Energy and Environmental

Policy Center, recognizing the contribution that this research could make to assessments of the Energy Security Act program, supported preparation of this book.

As we move further into the 1980s, the economic and political environment that fostered the Energy Security Act program has changed in some important ways and remained the same in others. The nation may reconsider—if not the existence of, at least the level, budgetary commitment, form, and objectives for—special government investments to promote initial commercial synthetic fuel projects. The various audiences for this book may interact in any such reassessment as well as in the implementation of the ongoing Energy Security Act program.

Too much depends on individual judgments of key probabilities, parameters, and effects for any book to provide many definitive assessments. However, by squarely, critically, and to some extnet quantitatively addressing the paradoxes and contrasts of initial synthetic fuels development, this book seeks to contribute to a more informed and balanced discussion of an important and potentially large national investment. Moreover, the book seeks to articulate an economic and strategic framework for national investments to promote initial private synfuels projects. Such a framework may help build a stable consensus for prudently, but cooperatively, proceeding with the major, multi-decade commitments required of private industry, governments, and the nation as a whole.

James K. Harlan
Washington, D.C.
December 1981

ACKNOWLEDGMENTS

This book owes much to the support, assistance, and guidance of a number of organizations and persons. Given the many individuals deserving recognition and thanks, any set of acknowledgements is likely to omit some who have helped along the road to this book. Recognizing this risk and apologizing in advance, important contributors to my efforts are noted below. Despite the range of assistance and commentary received, the descriptions, results, and conclusions presented in this book represent only this author's views and interpretations. They do not necessarily reflect the views of any other individual or organization. Of course, I bear full responsibility for the remaining errors of comission, omission, and misinterpretation.

It is most fitting that this book is being published as a report of Harvard's Energy and Environmental Policy Center (EEPC). As it has for many subjects, the EEPC provided an academically rigorous, but practically oriented, environment for the original investigations upon which this book is based. Professor William W. Hogan, the Center's Director, was a formative force in this effort. He provided important opportunity, resources, and substantive advice for the application of decision analysis to the complex public investment questions posed by commercial synthetic fuels development. Two other members of the EEPC's Harvard Synfuels Group—Ludo Van der Heyden (now at Yale) and Paul Leiby—made major contributions to the structuring and implementation of the decision analysis model. Henry Lee, Executive Director of the EEPC, provided valuable comments on the original research as well as enthusiastic encouragement and financial support for preparation of the manuscript. The Center's administrative staff—including Joan Curhan, Dawn Bursk, and Laura Kelly—graciously provided facilities and clerical assistance.

Significant parts of this book are based on work originally performed as a special project for the Department of Energy's Office of Policy and Evaluation. Alvin L. Alm encouraged my efforts to address basic questions regarding synfuels development both in his DOE management capacity and afterwards from his more academic viewpoint. John M. Deutch was similarly supportive. A number of present and former DOE officials provided opportunity and encouragement for the original analytical efforts from which this book is derived. They include: William J. Silvey, David Bodde, A. Denny Ellerman, Nick Timenes, Stuart W. Ray, and Bruce Robinson.

Development of this book has benefited greatly from the critical, but supportive, review and commentary of professional friends and colleagues in both Boston and Washington. Tom Schelling, Henry Jacoby, and Howard Raiffa reviewed an early manuscript; each suggested helpful refinements and improvements. Several policy practicioners in Washington were also generous with their scarce time. George R. Hall reviewed several interim drafts and provided comments which were invariably insightful and useful. Jack M. Appleman assisted greatly with substantive comments as well as the hospitality of his home and office. Others who provided important encouragement and opportunities to test ideas included Art Ingberman, Larry Linden, Edward Merrow, Robert Shelton, and Richard Richels.

The development of this book has also benefited from the industrial experience that several people in the private sector shared in interviews. Of the many people who were generous with their valuable time, William Henze—Vice President for Synthetic Fuels at The Badger Company in Cambridge, Massachusetts—deserves particular acknowledgement. Mr. Henze was extraordinarily generous with his time and insights on the design and learning process for chemical process technologies. Other individuals who graciously provided useful interview commentary included Dana Lee, Vince Kavelick, Edward McIves, and Maurice L. Mosier.

Editorial assistance for preparation of the book manuscript was superbly provided by Betsy Gilligan. Ms. Gilligan quickly mastered the technical content and key messages of a difficult draft manuscript intended for diverse audiences. She provided and negotiated both major and minor editorial suggestions that substantially improved the manuscript. Her remarkable commitment, diligence, and friendly enthusiasm in the editorial effort made a difficult task much quicker, easier, and more pleasant. The remaining inadequacies and obscurities of this volume's expression probably result from this author's failure to accept some of Ms. Gilligan's editorial suggestions.

The manuscript for this book was prepared during the summer of 1981 with editorial expenses provided by Harvard's Energy and Environment Policy Center. That assistance is gratefully acknowledged. Occasional clerical support for manuscript preparation was efficiently provided by Cyndi Flores and Karen Loveridge.

The long road to this book has been professionally satisfying. It has also been personally satisfying, bearable, and even enjoyable thanks to friends and loved ones. Many friends—some of whom have been noted above—assisted my efforts via the hospitality of their homes during my trips between Boston and Washington. My family, though sometimes puzzled by the substance and context for my work, has been consistently supportive. However, my deepest appreciation extends to Mary Ellen Bopp, whose competence, grace, sacrifice, and gentle insistence provided important opportunity and encouragement throughout the course of this work, especially through the restructurings of the intense summer of 1981.

J. K. H.

STARTING WITH SYNFUELS

1 INTRODUCTION AND OVERVIEW

THE PARADOX OF CERTAIN POTENTIAL AND UNCERTAIN ROLE

Synthetic fuels present a paradox. Although they can certainly be produced in large quantities, the role for the synfuels option is far from certain.

Certainties and Uncertainties

Enormous hydrocarbon resources lie beneath the United States in the form of coal and oil shale. Technologies to transform these resources into "synthetic" substitutes for naturally-occurring oil and gas have been used and under development for decades. The potential for combining natural resources and capital investment to produce substantial quantities of synthetic fuels is undeniable. A single synthetic fuels plant is certain to require capital investment of several billion dollars. Because of the lack of practical production experience, however, the actual investment cost, production rate, environmental impacts, and effective per barrel product cost remain uncertain.

Surges of Interest

Despite these uncertainties, the undeniable potential of synthetic fuels has motivated their high profile role in many recent proposals for reducing oil imports. Even now, the United States stands poised to undertake, with subsidies if necessary, a major program of investment in commercial synthetic fuel production facilities. Following the oil embargo of 1973-74, synthetic fuels production was a major component of the aborted Project Independence blueprint. In 1975, the Ford Administration proposed a major program for achieving one million barrels per day of synfuel production by 1985 (Synfuels Interagency Task Force 1975). However, legislation to implement the Ford administration proposals was narrowly defeated in the Congress in 1976.[1] Through the 1970s, no major program to support commercial synthetic fuel production was adopted.

In 1979, a second oil supply disruption—associated with turmoil in Iran—renewed intense interest in alternatives to imported oil. Again, synthetic fuels figured prominently in proposals for reducing oil imports. In June 1979, with gasoline lines spurring cries for action, the U.S. House of Representatives overwhelmingly passed legislation calling for commercial synthetic fuels production reaching 500,000 barrels per day by 1985 and two million barrels per day by 1990.[2] In July 1979, the Carter Administration proposed a major program to reduce oil imports, which included a synthetic fuels component calling for 1.5 million barrels per day production by 1990.[3] At the same time, the Senate Energy Committee was considering its own bill for oil import reductions, which included several synthetic fuels provisions. Through the ensuing legislative process, these various proposals were coalesced to yield, in June 1980, the Energy Security Act of 1980.

The Energy Security Act addresses a number of alternatives for reducing oil imports but emphasizes commercial synthetic fuels production.[4] It establishes national synthetic fuel production goals of 500,000 barrels per day by 1987 and two million barrels per day by 1992 and authorizes monies for economic and financial incentives required to move toward these goals. The new, publically chartered and funded corporation it creates—the United States Synthetic Fuels Corporation (SFC)—has the mandate to plan and administer support for initial commercial synthetic fuel production efforts.

ASSESSING THE SOCIAL INVESTMENT IN SYNTHETIC FUELS

The intensified interest in reducing oil imports which fostered passage of the Energy Security Act stems primarily from increasing concern over the economic costs and national security liabilities associated with the international oil market. The social costs of oil imports have been recognized to exceed their market price. These added costs contribute to an oil import "premium" (Department of Energy 1980c; Hogan 1980; Schelling 1979).

Concern about the oil situation has encouraged enthusiastic support and substantial subsidies for domestic energy alternatives with costs substantially exceeding present or projected costs for imported energy. However, as many economists observe, oil imports, even with proper accounting for an oil import premium, may provide energy at a lower cost than many domestic alternatives (Joskow and Pindyck 1979; Schmalensee 1980; Jacoby et al. 1979). The oil import premium is not infinite. Efficient use of the nation's resources requires that less costly energy options be used before more costly alternatives. Moreover, an oil import premium does not justify placing special emphasis on any particular energy option such as synthetic fuels, nor does it provide guidance for assessing the appropriate volume of initial synfuel production.

Synthetic fuels production will require enormous investments of society's limited resources. These investments have opportunity costs as alternative investments are deferred in order to make resources available for synfuel development. In approaching such major investments, it is important to go beyond national security rhetoric and beyond general concern for the full costs of oil imports. Decisions should be predicated on careful assessments of the potential benefits and costs. This study contributes to such assessments.

Three Basic Questions

This study addresses three basic questions concerning initial synfuel production programs. The first is "Whether and why?" the United States should make special efforts to realize initial synfuels production experience. Part I of this study describes a taxonomy for potential benefits and costs of initial synfuels production experience that suggests possible answers to this question.

The second question is "How large?" an initial production program is worthwhile. Part II develops and implements a decision analysis framework and model to systematically assess this question. Numerous scenarios and varying assumptions for key uncertain parameters—as well as subjective expectations for our energy future—are examined in addressing this question.

The third question is "How to do better?" with an initial synfuels production program, regardless of its size. Part III focuses on the linkages between key sources of benefits and the nature of initial production efforts. It suggests program design issues and approaches for improving the cost-effectiveness of an initial synfuel production program.

"Whether and Why?" a Synfuels Program. The taxonomy of benefits and costs characterized in Part I of this study reflects the uncertainties and dynamics of the synfuels decision problem. Uncertain synfuel production costs combine with uncertain oil price trends to cloud the economic value of synfuels production. However, the demonstrated potential for rapid changes in the market price or security costs of imported oil could increase the costs to the nation of oil imports, leading to efforts to aggressively expand synfuel production capacity. Three broad categories of benefits from initial production experience may be defined.

First, *production benefits* (or costs) for each barrel of synfuel equal the difference between the costs of producing synfuels and the full cost of a barrel of imported oil. The cost to the nation of imported oil is the market price plus the additional costs reflected by an oil import premium.

Second, *information benefits* may be obtained because reduced uncertainty about synfuel production costs can improve future investment and policy decisions. The value of information is magnified by the long-term and capital-intensive character of many energy investment alternatives.

And third, *surge deployment benefits* may be realized as initial production experience improves the nation's capabilities to undertake an aggressive expansion of synthetic fuels production capacity. Surge deployment benefits are analogous to insurance payoffs against sudden crisis events. Initial production experience acts as insurance allowing a surge deployment effort to achieve "lower cost" plants (*learning*) and "more plants" (*enhanced deployability*). Surge deployment benefits recognize the potential for rapid change in energy markets and the social benefits of an improved national capability to respond to our uncertain energy future.

These important sources of benefits and costs from initial synthetic fuels production are further defined and characterized in Part I. Parametric estimates are presented and show potential benefits and costs to be about the same size.

However, important sources of benefits and costs have contrasting implications for the appropriate size of an initial production program. *Production benefits or costs* vary in direct proportion to the amount of initial production. Values are uncertain because they depend on uncertain outcomes for synfuel production costs and oil price trajectories. Either costs or benefits relative to oil are possible.

Uncertainty regarding the basic cost level for synfuels will be clarified by the first few production facilities. Accordingly, *information benefits* can be obtained with a very small initial production effort.

Surge deployment benefits may be separated into two categories, each having contrasting implications for program size. Since learning with synthetic fuels technology is largely a sequential process, *learning* benefits do not increase greatly for larger volume initial production programs. In contrast, building infrastructure to increase future deployment capability may exhibit a "critical mass" relationship to program size. *Deployability* benefits may exhibit increasing returns to program size. However, larger initial production programs incur added costs as excess demands for some inputs lead to price increases, bottlenecks, delays, and other inefficiencies which increase the costs for all plants built. These added costs from *factor cost inflation* are much greater for larger than for smaller initial production programs.

A qualitative and quantitative review of these competing benefits and costs suggests that the answer to the "Whether and why?" question is: Yes, the United States should seek—and even subsidize—some initial synfuels production experience because the improved information and capabilities from that experience may provide worthwhile insurance payoffs in some energy futures.

"How Large?" a Synfuels Program. Part II considers the issue of "How large?" an initial synfuels production effort is appropriate, or

"How much?" insurance the nation should purchase. The analysis is based on the framework for benefits and costs developed in Part I. Benefits from information and improved surge deployment capabilities must be weighed against costs for production subsidies for various size initial production programs.

Surge deployment benefits are like insurance "payoffs" that accrue only when some extreme event motivates an aggressive synfuels deployment effort. Costs for production subsidies are like "insurance premiums" necessary to obtain insurance protection (improved surge deployment capabilities), which one hopes will not have to be used. The net comparison of costs and benefits depends on one's view of key uncertainties, such as synfuel production costs, and the likelihood that an "insurance event" will require that surge deployment capabilities be utilized. In some scenarios, there are large subsidy costs and few benefits; in others, few costs and substantial benefits. Given the uncertainties and the insurance character of key benefits, it is necessary to examine many possible outcomes and scenarios. Extreme scenarios should be weighed along with more likely scenarios to assess expected net benefits (or costs). Scenarios more likely to occur should be weighted more heavily than lower probability scenarios.

Part II develops a decision analysis framework and model that addresses the "How large?" question. The decision analysis structure allows the inherent uncertainties for many key parameters of the synfuels insurance problem to be separately considered and characterized, yet systematically integrated.

Chapter 4 outlines the basic structure of the decision analysis model, Chapter 5 describes how the model works, and Chapter 6 presents and interprets the decision analysis results for a set of base case assumptions and a large number of sensitivity cases. Key sensitivities examined include long-run oil price trends, synfuel production costs, the probabilities and descriptions for oil price jump events, the additional costs for larger programs due to factor cost inflation, and the importance of infrastructure development.

Decision analysis results show that a "moderate" initial program (which provides learning and builds some infrastructure) is a good choice for a wide range of outcomes for these key sensitivities. However, the results presented in Chapter 6 are not intended to be definitive. Other specifications for parameters and probabilities could change results. The decision analysis approach and model are presented as tools that can be used to help illuminate appropriate choices for the complex issues, tradeoffs, and judgments that are part of designing and implementing an initial synfuel production effort.

"How to Do Better?" Part III of this study shows that the design and implementation of an initial synfuels production effort, regardless of its size, can be more cost-effective if important sources of benefits are accented. The surge deployment benefits of *learning* and *enhanced deployability* provide special guidance for program design. Assessment of the "linkages" between initial production experience and improved surge deployment capabilities can suggest issues and approaches for more effectively realizing the potential benefits of initial synfuels production experience.

Chapter 8 considers these linkages for learning benefits. Program design attention to learning effects argues for production phasing that responds to the sequential nature of the learning process and provides adequate time to exercise opportunities for sequential learning. Also, learning *which* technologies to employ in possible surge deployment efforts ("selection learning") requires attention now to the information that will be needed to make improved technology comparisons and investment choices in the future. Smaller initial commercial projects could advance learning more effectively than larger projects. When emphasizing learning benefits, information and technology improvements are a more important output of the production activities than the actual barrels of synfuel products.

Chapter 9 considers linkages between initial production activities and infrastructure development for *enhanced deployment* capabilities. For strategic insurance investments in improved surge deployment capabilities, key deployment constraints are those that can be relaxed only via long lead-time investments before a maximum deployment effort becomes necessary.

Many constraints can be relaxed with effort after aggressive deployment is clearly warranted. However, other constraints can be relaxed only via practial experience requiring substantial time. Such long lead-time constraints limit existing deployment capabilities and the practicable size for an initial production phase. They also provide a useful focus for strategic investments to build key elements of the synfuels industrial infrastructure as insurance for future surge deployment scenarios. For example, an important existing constraint for synfuels deployability stems from limitations on the number of architect, engineering, and construction (AE&C) firms that are experienced and well-qualified to undertake a multi-billion dollar synfuels project. Future deployment capability may be enhanced if an initial production program is structured and implemented to assure that some new AE&C capability is developed or trained via initial production efforts.

Conclusions for the SFC Program. The discussion in the first three parts of this study is purposely generic. Benefits, costs, and program design issues can be assessed in general without being tied to any particular program for initial synfuels production. Chapter 10 moves away from this generic presentation to relate the conclusions of this analysis to a major practical implementation context—the program for initial synfuels production authorized by the Energy Security Act of 1980 (ESA), to be administered by the United States Synthetic Fuels Corporation (SFC). By addressing "Whether and why?" an initial synfuels production program is warranted, Part I articulates a statement of key elements of the SFC mission. The analysis in Part II helps characterize "How large?" the SFC program should be. The discussion in Part III suggests program design provisions and issues that the SFC should consider as part of cost-effectively performing its important mission.

The conclusions of this analysis give mixed reviews to the program defined by the Energy Security Act of 1980. The importance of learning, infrastructure development, and a phased approach are recognized by the act, but tensions between strategic capability purposes and the act's ambitious production goals remain unresolved. Unless clarified by congressional review, the legislation's apparent emphasis on production volume as a primary objective and measure of accomplishment could encourage the SFC to overemphasize simple production volume for the many tradeoffs and discretionary choices it must make. Part II's analysis shows that the size of the SFC program's first phase is justifiable. It may well strike a good balance between learning, infrastructure development, and excessive factor cost inflation. However, Part III shows that the timing for planning and commitment to ambitious second phase production levels is premature. In order to benefit from the important information and technology improvements that are likely to result from first phase projects, the timing for both the second phase decision point and production goal should be shifted to about three years later.

Chapter 11 concludes this study by summarizing eight major lessons from this analysis for the SFC program in particular and for initial synfuels production efforts in general.

A GOVERNMENT PROGRAM AND THE FREE MARKET

This study considers benefits and costs for the nation as a whole. It does not distinguish between those benefits that can be captured by private firms through marketplace transactions from those that accrue to society at large.[5] Private firms are part of society. Some benefits to the nation provide profit opportunities to individual private firms. However, other benefits are "externalities" that improve the posture of the nation as a whole without improving the balance sheet for an individual project.[6] The use of "social" benefits avoids the need to distinguish between benefits and costs that are included and excluded from the assessments of individual private firms.

The use of "social" benefits and costs also necessitates careful interpretation of the results of this study. Although the study shows that the nation will be better off with some synthetic fuels production experience, the presence of net benefits for the nation as a whole does not imply that a special government program to promote initial production is necessary or appropriate. A portion of the total benefits of pioneering synfuels projects can be captured by the private firms sponsoring such projects. Market opportunities and private incentives alone *could* yield substantial initial synthetic fuel production without special government incentives or program efforts.

However, this study argues that many of the benefits to the nation are externalities that do not benefit the private firms undertaking pioneering projects. Accordingly, free market incentives and private decisions alone may not lead to the levels and forms of initial production experience that provide an appropriate strategic posture for the nation as a whole. Some special government program for initial synthetic fuels production may be warranted.

In general, we should recognize the value of marketplace incentives and unfettered private decisionmaking for allocating resources and implementing projects. Special government action may be appropriate, however, where free market incentives and individual private decisions alone are unlikely to address adequately the interests of the nation as a whole.

EMPHASIS AND LIMITATIONS OF THIS STUDY

Generic, Coal-Based Synthetic Fuels

"Synfuels" is a general term that commonly includes a wide range of resource bases, technologies, and products. However, the direct focus of this analysis is limited to *coal-based* synthetic fuels. Moreover, technology for the production of coal-based synfuels is treated primarily in a generic fashion without attention to alternative technological methods or product slates. This very aggregate depiction of the coal-based synfuels option suppresses some details, issues, and alternatives. Synfuels technologies, products, and costs are sufficiently similar, however, that major issues for assessing the synfuels insurance investment can be addressed appropriately with such a generic approach.

Important differences in the economics and issues between different synfuels do exist. In particular, one reason for excluding oil shale from this analysis is that oil shale from rich western deposits generally is expected to have lower costs than coal-based synfuels. Given the oil price increases of 1979–80, adequate development incentives may exist from private sector market opportunities alone. The framework for benefits and costs underlying this study emphasizes the insurance value of initial commercial experience. Special insurance investments are particularly important for technologies where the production costs are too high to allow the private sector to move ahead without subsidies now, but low enough for the technology to play an important and economic role for some plausible, albeit unhappy, developments in world oil markets.

The approach and analysis in this study have obvious analogies to other resources and technologies. However, in this study, the terms "synthetic fuels" or "synfuels" should be understood to mean technologies and products based on coal as a feedstock.

Two Issues Not Addressed

The agenda of issues relevant to the evaluation and design of initial synthetic fuels production efforts is clearly extensive. Given the conceptual emphasis of this study, many issues are not addressed in detail. Often, parametric assumptions and sensitivity analyses are used to assess the relative importance of some issues. For example, the study does not attempt to predict synfuel production costs. Rather, the discussion in Chapter 3 characterizes a reasonable range of uncertainty for synfuel costs. Parametric estimates over this range are then used to assess the size of potential production subsidies. Particularly in Part III, program design issues requiring extensive analysis are identified with little further comment. Such issues are noted in order to suggest priority issues and useful approaches for further assessments.

Inevitably, many issues are not adequately addressed in this study. Two deserve particular mention. First is the level of private investment likely to occur in response to market developments alone. The expected level of private investment in synthetic fuels without special government incentives is a key determinant for the role for government programs. Prospects for unsubsidized private investment in commercial synfuel production have changed as markedly as oil prices in recent years. Government programmatic initiatives further complicate assessments of what the private sector will do or might have done in response to market developments alone. Since this study does not describe the response of private investors to market incentives alone, results bearing on the appropriate level of initial synthetic fuel production efforts for the nation as a whole are *not* necessarily a recommendation for a correspondingly large government program.

Second, this study only briefly considers the environmental costs and concerns or the effects of synthetic fuels development on the socioeconomic fabric of particular localities, states, and regions. These costs and issues should be recognized and included in policy formulation as well as in program design and implementation. Fortunately, other studies are giving extensive attention to those issues (For example: Office of Technology Assessment 1980; Energy Research and Development Administration 1977; Department of Energy 1981a; Synfuels Interagency Task Force 1975 vol. 4; Seidman 1980). Integration of those assessments into formulation of initial synfuels production efforts is an important complement to this study.

How This Study Differs From Other Synfuels Studies

Any program of the size and cost of an initial synfuel production effort generates attention, study, and controversy. Advocates come forth and critics arise. The profile of synthetic fuels among our energy alternatives and in government program initiatives has generated many studies and much commentary.

Many synfuels studies describe synfuels technologies, production costs, environmental impacts, and other issues. Some studies primarily describe technologies. Other synfuel studies assume a production goal and examine costs, technology options, material and labor requirements, and other issues associated with meeting the goal. Still other studies compare synfuels production with other ways of reducing oil imports.[7]

This study assesses several key questions for a initial synfuels production effort without describing the synfuel technologies, environmental effects, or other issues in detail. For the questions addressed in this study, the details of technologies, flowsheets, and effluents are not as important as a general understanding of production costs and strategic issues.

Unlike many synfuels studies, this study does not assume the inevitability of a continuing synfuels industry or a particular production growth trajectory. The uncertainties and dynamics that could accelerate and accentuate—or defer and eliminate—any major long-run role for synthetic fuels are recognized and reflected in this analysis. Rather than projecting or assuming a production schedule decades into the future, this study addresses the question of how much *initial* production experience is appropriate for the nation, given the uncertainties that we face. Further, this study does not attempt to compare synfuels to other major energy options such as conservation, solar, nuclear, and so forth.

Many studies assume that some government synfuels program is warranted and examine alternative structures and incentives for such a program (Booz, Allen, and Hamilton 1979; Department of Energy 1979a; Sobotka & Co., Inc. 1978). This study, however, does not assume that a government program of any particular scope or type is necessary or appropriate. Moreover, it does not address the wide range of alternative financial instruments, institutional arrangements, or other incentives that government could offer to encourage additional private investment in initial synthetic fuels projects. This study focuses on the benefits, costs, and appropriate form for initial synfuels production experience. It does not describe the tools or incentives required to obtain that experience.

A Spectrum of Views on Synfuels. Synfuels advocates often assert an imperative to reduce oil imports, citing high prices and the insecurity of imported oil supplies. Synfuels are promoted as a sure and effective way to reduce oil imports. Advocates often fail, however, to explain exactly how synfuels will increase our energy security, how much we should be willing to pay to avoid a barrel of imports, or why synfuels merit special attention among the many alternatives for reducing oil imports.

At the other end of a spectrum, many economists argue forcefully against any special efforts to promote synfuels production. They cite potential efficiency losses as well as the difficulties and distortions of a government program (Schmalensee 1980; Joskow and Pindyck 1979; Jacoby et al. 1979). They note the availability of many lower cost options for reducing the level of oil imports or, more importantly, for reducing the nation's vulnerability to oil supply disruptions. Market processes are advanced as the best way to allocate resources among alternatives for reducing oil imports. Such simple descriptions of marketplace efficiency, however, often suppress important dynamic issues such as infrastructure development, technology improvement, and the potential for rapid change in the market price or security costs of oil imports.

This study occupies some middle ground in this wide spectrum. It is based on the premise that the assessment and design of an effort as potentially important and costly as an initial synthetic fuel production program should not be guided by simple, uncritical assertions either of national security imperatives or of the general adequacy of the free market and unfettered private decisionmaking for the appropriate allocation of resources. Its analysis is based on economically calculable costs and benefits. It recognizes the value of market processes but also incorporates the important dynamics of potential rapid change in energy markets and the time required to develop improved capability and infrastructure for important energy options. A key precept of the analysis is economic rationality rather than political imperative.

Assessing the Synfuels Insurance Investment

Given limited resources and competing national concerns, the substantial investments required for initial synfuels production should be carefully considered. The three parts of this study address basic questions for national investments in initial synfuels production experience:

1. "Whether and why?" we should proceed;
2. "How large?" an investment we should make; and
3. "How to do better?" with whatever investments we decide upon.

Systematic assessment of these questions quantifies and illuminates important issues for the design of a prudent, well-grounded initial synthetic fuel production program. The investments and decisions are large. Stable, long-term commitments must be made. The Congress, the administration, the Synthetic Fuel Corporation, private industry, and the public at large all face important choices regarding investments with the synfuels option. The perspective presented in this study seeks to contribute to improved discussion, design, and implementation of those major decisions and investments.

NOTES

1. For background, see the legislative history associated with H.R. 12112, 94th Cong., 2nd Sess. (1976).
2. The Moorehead synthetic fuels bill was part of an extension of the Defense Production Act. The legislation was developed by the House Banking Committee's Economic Stabilization Subcommittee. It passed the House of Representatives by a vote of 368-25 on June 26, 1979.
3. The Carter administration proposal included biomass, heavy oil, and unconventional natural gas in the definition of "synthetic fuels." The 1.5 MMBD goal noted includes estimates for coal synthetics (1.0 to 1.5 MMBD) and oil shale (0.4 MMBD) (White House 1979).
4. See *Energy Security Act of 1980*, Pub. L. No. 96-294, enacted June 30, 1980.
5. Germany, Japan, and other international allies will benefit from improved information and capabilities with synfuels; however, "society" or "the nation" refers primarily to the United States.
6. "Externalities" is an economic term referring to benefits or costs that are not included within (i.e., internal to) the calculation of benefits, costs, and trade-offs of a particular decisionmaker. For more information, see most basic economics texts, including Baumol (1972).
7. For studies emphasizing technology descriptions, see Hottel and Howard (1971), Department of Energy (1978), and National Research Council (1977), among many others. Studies examining the implications of an assumed production goal include Bechtel National, Inc. (1979), Gallagher, et al. (1976), Department of Energy (1979a), and Energy Research and Development Administration (1976a). Studies addressing benefits and costs include Synfuels Interagency Task Force (1975) and ICF Incorporated (1979b).

I WHETHER AND WHY A SYNFUELS PROGRAM?

Part I of this study addresses the fundamental issues of "Whether and why?" the United States should make special efforts to promote initial production of synthetic fuels. Resources directed toward synthetic fuel production force curtailment of some other valued activity, product, or effort. We should carefully assess how usefully our limited resources are allocated.

Part I considers the special circumstances of energy markets in general and synthetic fuels in particular that could make the free market inadequately serve the national interest. Programs that overstep the normally adequate and efficient resource allocation mechanism of private firms and consumers responding to market incentives deserve careful review. We should ask whether and why special government action is warranted.

Initial synthetic fuel production may well incur costs for subsidies and other incentives. Such costs should not be borne unless the expected benefits to the nation are correspondingly large. Important sources of benefits and costs should be identified and, where possible, quantitatively compared. The result of such cost-benefit comparisons should clarify whether, why, and under what circumstances an initial synfuels production program is worthwhile.

Part I develops a framework for assessing the benefits and costs of initial synthetic fuel production efforts: Chapter 2 develops a comprehensive framework for key sources of benefits, and Chapter 3 a corresponding framework for costs. Both chapters present qualitative reviews as well as quantitative estimates of the potential magnitude of benefits or costs.

The analysis shows that net benefits or costs depend on particular assumptions for uncertain parameters or scenarios. In some cases, benefits are small and costs large; in other scenarios, benefits, substantially exceed costs. Benefits and costs can also be about the same size. It is not a simple situation. Special investments in initial synthetic fuels production are neither clearly a good or a bad idea.

The framework shows that there are some good answers for the "Whether and why?" question that apply to synfuels in particular. Moreover, the concepts and quantitative assessments help illuminate the analysis described in Part II of this study, which addresses the question of "How large?" an initial synthetic fuel production program is appropriate. Furthermore, the framework for benefits and costs suggests important program purposes and design issues, providing the basis for the assessment of the question "How to do better?" which is addressed in Part III.

2 BENEFITS FROM INITIAL SYNFUELS PRODUCTION
A Framework and Assessment

INTRODUCTION

A well-defined framework for the benefits of initial synthetic fuel production efforts can improve investment and design decisions for initial synthetic fuel production efforts.

This chapter develops a comprehensive taxonomy for the potential benefits of initial experience with synthetic fuels production. It is structured around three broad classes of possible benefits, each class having a key distinguishing attribute.

First, *production benefits* derive from the physical contribution that synthetic fuel products make to energy supplies and their associated impact on energy markets. Their key attribute is that they are linked directly to each unit of production. Consequently, benefits are approximately directly proportional to initial production volume.

Second, *information benefits* stem from the improved public and private investment and resource allocation decisions that are possible with better information about our major energy options. A key attribute of information benefits is that they arise from basic experience with synthetic fuels production and thus are not linked to each unit of production. Moreover, the magnitude of information benefits is not especially sensitive to future oil market scenarios.

Third, *surge deployment benefits* arise from improved capabilities with the synthetic fuels option that are fostered by initial practical experience. These improved capabilities are like insurance "payoffs". Their key attribute is that they become important primarily during a surge deployment scenario where intensive efforts to realize quantum increases in synthetic fuels production capacity become necessary.

As with most categorizations, some sources of benefits may not fall neatly into a particular category. However, it is useful to distinguish between benefits that are linked to each unit of production, benefits that accrue from general experience, and benefits that are important primarily during special emergency event scenarios.

This taxonomy of benefits attempts to be comprehensive in order to cover the major rationale for programs to support initial commercial-scale synthetic fuel production. Additional qualitative discussion and some parametric quantification are presented for each of the taxonomy's three basic categories of benefits.

PRODUCTION BENEFITS

Production benefits are those values directly associated with each unit of synthetic fuel product. Two major sources of potential production benefits are readily identified: the market value of synfuel products and the oil import premium.

Production values represent the market value of synfuel products compared to their production costs accumulated over the lifecycle of initial production facilities. Although these comparisons of synfuel costs and oil prices could show net benefits or net costs, this discussion assumes that initial production projects will incur net costs and require production subsidies. Accordingly, the market value of synfuel products is not discussed as a program benefit but rather is considered in the taxonomy and assessment for synfuel program costs in the next chapter (Chapter 3).

Each barrel of synfuel production also yields production benefits associated with the oil import premium. By eliminating a barrel of imported oil, each barrel (oil equivalent) of synfuel production avoids the additional social costs, over and above the market price, of imported oil supplies. These additional costs to the nation are represented by the "oil import premium." Elements of the oil import premium include a number of economic effects as well as the national security liabilities associated with the worldwide trade in oil. The per barrel magnitude of this premium has important implications for policy and cost-benefit assessments. The oil import premium defines the amount above the expected market price for oil we should be willing to pay for domestic alternatives to imported oil.

Major components of the oil import premium and some recent assessments of the possible magnitude of the premium are described further in the next section. The range of potential values for an oil import premium is wide, but such a premium is not clearly so large as to make the full social cost of a barrel of imported oil necessarily exceed the uncertain the potentially high production costs for coal-based synthetic fuels. (These costs are characterized in Chapter 3.) Moreover, a high oil import premium neither provides guidance to program design nor supplies any rationale for special emphasis on synthetic fuels as compared to the many other alternatives for reducing oil imports.

National security concerns—a high profile rationale for oil import reductions in general and for synthetic fuels production in particular—are difficult to cast into the same economic framework as other elements of the oil import premium. Unlike most production benefits, some energy security effects may not necessarily be linked to each unit of production. Accordingly, a separate section summarizes an assessment of the national security benefits of initial synthetic fuels production.

Benefits from an Oil Import Premium

The concept of an oil import premium is becoming widely accepted. There are a growing number of qualitative and quantitative assessments of the benefits of eliminating a barrel of imported oil over and above the market price of the oil displaced.[1] The elements of an oil import premium vary depending upon the assessment but generally include:

1. An *oil pricing* benefit whereby lower demand for oil imports reduces pressure on world market supplies and leads to lower prices. The price savings obtained via reduced demand at the margin are obtained over the total volume of barrels purchased.

When these savings are credited to the source of import demand reduction, substantial per barrel benefits (or "premiums") may be estimated.

2. *Domestic economic* benefits include improved balance of payments arising from reduced dollar outflows for oil purchases, reduced inflation when price increases for petroleum products occur, and expanded domestic employment. When such potential economic benefits are credited to the source of oil import reduction, the per barrel premium increases.

3. *Energy security* benefits arise from the reduced role of unreliable oil supplies in domestic and free world economies. A reduced role for oil imports will reduce the microeconomic and macroeconomic costs of supply interruptions or price increases. Reduced economic costs should reduce the associated political leverage available from the threat of supply or price manipulations. Reduced oil import levels should also reduce the necessary size, stringency, and cost of various measures, such as the strategic petroleum reserve, designed to deal with disruptions in energy markets.

Of the several recent analyses that have attempted to quantify the magnitude of various contributions to an import premium from these general elements, an assessment by William Hogan (1980) in *Energy and Security*, is particularly clear, systematic, and comprehensive. Hogan presents carefully derived quantitative estimates for major components of an oil import premium, including inflation and balance of payments effects, oil price effects stemming from potential monopsony power of consumers, and vulnerability costs associated with the size and probability of oil import interruptions. He shows that plausible differences in opinion and assessments regarding parameter values, supply/demand responses, policy responses, and the probabilities associated with a particular world view could lead to an oil import premium range of from $4 to $40 per barrel for the United States alone. A similar assessment by the Department of Energy's Office of Policy and Evaluation (1980c: VI-22 − VI-25) suggests a range for the long-run oil import premium of about $3 to $10 per barrel, depending on particular expectations for disruptions in world oil markets and total levels of oil imports.

Although neither of these assessments addresses the benefits of increased domestic employment, those benefits are likely to be small both initially and over the long run. Many of the labor resources employed for synthetic fuels projects (or other energy projects) are presently fully employed and are likely to continue being fully employed in alternative economically productive activities. Synfuel plants require the services of engineers, managers, and skilled construction craftsmen. Unemployment or under-employment of such personnel is not presently a problem. The primary incremental employment impact of synfuels construction is likely to be for general unskilled or semi-skilled labor. Growth in many types of industrial activity also could provide such general employment.

The plausible range of values for an oil import premium is clearly substantial. The range of potential production benefits from initial synfuels production is correspondingly large since the net benefits depend strongly on the value assumed for the oil import premium. According to both the Hogan and DOE assessments, a range of values on the order of $10 per barrel is plausible, with values one-half or twice that amount being readily arguable. Such a range of uncertainty is similar to many ranges for synthetic fuel production costs. This analysis will not attempt further refinement or estimates of the oil import premium; rather, the production benefits will be reflected by an assumed value for the oil import premium.

For quantitative estimates, the benefits from an oil import premium can be calculated by multiplying synfuel quantities by the size of the premium and adding the result over the years of production. For net present value estimates, appropriate discounting should be applied to reflect the time value of expenditures and benefits. Alternatively, the value of the oil import premium can be reflected in the calculation of potential production subsidies by using an "effective" synfuel cost derived by subtracting the long-run oil import premium from the synfuel production cost. The import premium thus acts as a credit which offsets part of domestic energy production costs. Parametric assessments (developed in Chapter 3) show the implications for the amount of production subsidies of changing synthetic fuel cost levels. The same tabulations also can be used to consider the effect of changing assumptions for the value of an oil import premium.

The oil import premium has important, but ambiguous, implications for the net social costs or benefits associated with each barrel of synfuel production. High oil prices and/or low to moderate synfuel production costs could combine to make synfuel production economic or nearly economic in terms of market prices alone. If so, the oil import premium, depending on its assumed magnitude, would increase the benefits of synfuels projects. Even those projects that were not quite economic in market price terms could yield positive benefits to the nation.

Under other market conditions, however, the benefits of an oil import premium would not offset the subsidy costs associated with initial synfuel production. If oil prices were to stabilize at about $30 per barrel (1980$), production of coal-based synthetic fuels costing in the $50 to $70 per barrel range would require subsidies in the order of billions of dollars for each plant.[2] A realistic oil import premium may be small, or certainly not so large as to offset possibly large gaps between synfuel production costs and oil prices.

For each barrel of synfuel production, the oil import premium provides benefits that offset potential costs for production subsidies. (Production subsidies are discussed and estimated in Chapter 3.) Whether oil import premium benefits are adequate to offset production subsidy costs (if any) depends on the assumed value for the oil import premium as well as outcomes for uncertain synfuel production costs and long-run oil price trends.

However, it is also important to note that synfuels have no special claim to the benefits arising from an oil import premium. The benefits stem from reducing oil imports. Since any measure that reduces oil imports yields these benefits, net benefits to the nation are greatest if the lowest cost measures to realize oil import reductions are used. Even if the oil import premium is large, synfuels and government subsidy programs may not be the most cost-effective way to respond. Although oil import premium benefits may partially or completely offset the subsidy costs for initial synfuels production, they do not provide special rationale for particular emphasis on synthetic fuels as compared to any other of our many options for reducing oil imports.

Energy Security and Synthetic Fuels

Advocates of special government incentives for synthetic fuel production often invoke energy security as a high profile rationale (Committee for Economic Development 1979; Energy Security Act of 1980; Sec. 100). Energy security is a critical but difficult to quantify national objective. Accordingly, benefit-cost assessments should consider carefully exactly if and how a synthetic fuels program might enhance our energy security. The energy security costs of oil imports are already reflected in the general oil import premium discussed above. An assessment of synfuels and energy security should focus on whether synthetic fuels have any special energy security role beyond that contributed by any alternative for reducing oil imports. This section reviews major issues for synthetic fuels and energy security. It concludes that there are few special security-related justifications or program design rationale for synthetic fuel production efforts.

The Nature of the Energy Security Problem. The critical distinction between dependence and vulnerability bears importantly on the limited security role for synthetic fuels and many other capital intensive alternatives to imported oil. *Dependence* on a supply source need not lead to economic damage or *vulnerability* if, when a particular supply source is interrupted, alternative supplies can be readily substituted to maintain the economic activities for which the disrupted supplies were used (Deese and Nye 1980; Hogan 1980). Because many capital intensive alternatives to oil imports cannot be deployed quickly in response to supply interruptions, their energy security value is limited.

The energy problem is fundamentally an economic problem with two related, but distinguishable, dimensions:

1. A long-run dimension of *economic adjustment* to the various resource endowment and market power situations that lead to possibly high oil prices and to the associated transfers of wealth from countries that produce oil to those that consume it; and
2. A *security problem* derived from the potential for disruptions in oil markets which can result in rapid price increases and consequent damage to oil importing economies. Both the occurrence and the threat of disruptions cause security concerns.

Economic adjustment by itself is not a major *security* concern. Long-run adjustment of production and consumption activities to changing costs occurs in all markets. Facile economic adjustment is important to an efficient allocation of resources and a healthy economy, but the need to make such adjustments in response to gradually evolving energy markets is not primarily a security issue.

The security issue for oil markets stems from the demonstrated potential for major interruptions in supply combined with the minor, but pervasive, role that oil plays for many economic activities. Potential disruptions in oil markets have important security implications because capital-intensive energy consumption and production systems take time to adjust to rapid changes in a supply/demand/price balance. In the short-run, the system must adjust to any effective shortfall by price increases, reductions in energy use and economic output, and various macroeconomic adjustments.[3] These domestic economic losses combine with international wealth transfers to reduce significantly the welfare of consuming nations. As a result, disruptions or the threat of disruptions can yield international political leverage.

Three Approaches To Enhanced Energy Security. Energy security measures can reduce these liabilities by limiting the *magnitude, probability of use,* or *period of effectiveness* of the economic leverage associated with the potential to control or interrupt a substantial portion of world oil supplies.

Limiting the magnitude of disruption leverage is best accomplished in any particular timeframe by strategic supplies (such as a strategic petroleum reserve) or emergency demand restraint measures. Such measures serve to reduce the effective size of an interruption and the associated disruption price increases and economic losses (Hogan 1980). Economic policies, such as tariffs, also can retain in consuming countries part of the international wealth transfers that might be generated by the disruption price increases (Department of Energy 1980c).

Long-term reductions in the base level of oil import demand also reduce the economic impact of supply disruptions somewhat by reducing the profile of oil imports in an economy. However, such reductions in the basic level of oil imports have limited value in reducing immediate price increases and other impacts of a supply disruption. Synthetic fuels are a capital-intensive alternative requiring years to deploy. Like many other long-run alternatives for oil import reduction, they can do little to mitigate the security threat of a supply interruption.

General oil import reductions, however, do reduce pressures and perhaps tensions in world oil markets, possibly reducing the probability of supply cutoff. A loose world oil market also increases the costs to those who might instigate a supply interruption. For example, as excess production capacity increases, a country or group of countries must cut back their production much farther to cause a painful shortfall. Deeper production cutbacks increase the risks and costs (from foregone revenues) to the instigators of a disruption.

However, even if reduced oil imports were of significant value in reducing the costs of supply disruptions, the size of present and projected U.S., European, and Janapense dependence on OPEC and Persian Gulf oil supplies does not suggest any obviously meaningful or obtainable watershed target for oil import reductions. Even with zero U.S. oil imports, the economic interdependence of free world economies would cause oil supply disruptions to yield real economic and political damage to U.S. interests.

Limiting the "probability of use" or "period of effectiveness" of the oil leverage involves measures that might influence the attitudes and positions of the many actors in the complex arena of energy security and the Middle East. Actions to reduce the probability that disruptions will occur supplement measures for responding to events when they do occur. Security measures along these lines include foreign policy positions, military and other defense postures, and economic interdependence between producer and consumer states. For the long-run geopolitical concerns, energy security also could be enhanced by reducing the length of time any oil leverage could remain effective.

Energy Security Attributes of Major Alternatives. Over the long-run, some political dimensions of energy security may be determined by the attitudes of oil producers as well as of international allies and adversaries toward the varying energy postures of different nations or groups of nations. Various "attributes" or characteristics of long-run alternatives to oil may have different effects on these attitudes. These energy security attributes could differentiate particular long-run oil import reduction options from each other.

Relative economics is a key distinguishing attribute because it is fundamental to efficient resource allocation and long-run economic health. Other attributes, however, could differentiate the security role of oil import reduction alternatives. The performance of synthetic fuels with respect ot these potentially differentiating attributes is reviewed below.

Economics. Lower-cost alternatives to oil imports provide a more credible response to the use of oil leverage because the long-run economic cost of responding to price increases or disruption threats is lower. Although there is much uncertainty about cost, synthetic fuels may be relatively costly compared to other alternative energy sources. However, once they become economic, synthetic fuel production costs are likely to remain stable.

Resource Size. Larger potential resources bases may impact attitudes differently. In particular, energy options involving large resource bases may have greater impact on perceptions since they represent significant alternatives to the enormous amounts of relatively low-cost hydrocarbon resources of the Persian Gulf region. A single major alternative supply may have more impact on the perspectives of the oil cartel than a series of minor alternatives. For example, geopressurized methane is likely to be more important than coal seam methane, and Venezuelan heavy oil deposits more important than outer continental shelf oil.

The enormous size of the coal and oil shale resource base makes synthetic fuels a major U.S. energy option that should not be ignored by allies and adversaries considering the long-run U.S. energy supply posture and U.S. options for responding to developments in oil markets. Since coal is more geographically dispersed than oil shale, its accessible resource size and production levels may exceed those for oil shale. Although the effective size of the shale oil resource base is enormous, the geographic concentration and location of the richer resources may limit maximum production levels. The size of the resource base is a security attribute that favors synthetic fuels, especially coal-based synthetics.

Expandability. The rate at which major alternatives can substitute for imports may affect strategic perceptions and attitudes toward consuming nations' ability to respond to developments with the supply or price of oil. For example, alternatives that are readily deployable, even if relatively expensive, could mitigate aggressive attitudes on the part of those who control oil or who might move to control supplies for political ends. For example, ready alternatives to Middle East oil would reduce the expected geopolitical benefits to the Soviet Union of moves to control that oil.

Once a sufficient base of experience and infrastructure for an industry is established, synthetic fuels may have security advantages with respect to this attribute. Government regulatory policies could also promote rapid expansion potential. Although synthetic fuels cannot directly mitigate short-term disruptions, they represent a long-run option that could reduce the period or time over which political leverage from oil could remain effective.

Certainty. Import reduction options that are relatively visible, tangible, and certain add credibility to U.S. response capabilities. Such credibility may affect the attitude of those who might consider using or obtaining the political leverage associated with the control of oil supplies. Synthetic fuels production, though perhaps more expensive than the range of less visible individual actions that would achieve an equivalent amount of oil conservation, seems more tangible and more controllable by unitary social decisions. Such relatively certain options may have a stronger impact on attitudes (Weinberg 1979). However, the apparent certainty for synthetic fuels should not be overstated. The large size and centralized production that contributes to the apparent certainty for synthetic fuels also creates risks by increasing the potential for difficulties with an individual project to delay of major units of production.

Energy Security Concerns Provide Little Special Rationale or Guidance for Synfuels Programs. An energy security role for synthetic fuels production could stem either from the physical contribution of barrels of secure supply or, possibly, from the impact of energy options on the attitudes of those who might manipulate oil supplies to political ends. The evidence is speculative and mixed for any special role for synthetic fuels in favorably affecting those attitudes.

There appears to be little reason to differentiate among alternative long-run oil import reduction options other than on the basis of economics. Any special security premium for actions that reduce oil imports should apply uniformly to all alternatives. Those options that are least costly for any given time period should be preferred.

Even if oil import reductions via synthetic fuels could substantially improve our energy security, assessments of the oil import dependence of the United States alone or combined with its allies in Europe and Japan does not suggest any feasible quantity of synthetic fuel production that might yield a quantum improvement in the energy security posture of the United States and its allies.

Overall, national security concepts do not provide much rationale for special emphasis on synthetic fuels, nor do our general energy security concerns provide much guidance for design of initial synthetic fuel production efforts. How much synfuels production will make us secure enough? In any event, perceptions and attitudes toward synthetic fuel costs and capabilities may have far greater strategic security significance than the physical barrels of production. Evaluation and design of initial synthetic fuel production efforts should be related to other concerns, benefits, and purposes.

INFORMATION BENEFITS FROM INITIAL SYNTHETIC FUELS PRODUCTION EXPERIENCE

Practical experience with synthetic fuels production will provide useful information. The benefits from this information are not directly associated with each barrel of production or dependent on future oil price scenarios. Rather, the basic experience should clarify many highly uncertain issues for the synthetic fuels option. The experience will help "put synthetic fuels into perspective for our national energy future" (U.S. Senate 1979:129). Information from initial production experience should improve future decisions in many areas, providing benefits that do not depend on the particular outcomes of the experience or the future scenarios realized.

This section presents a qualitative discussion of two major ways in which information from initial production experience could yield benefits to the nation: (1) improved public and private policy and investment decisions and (2) the potential impact of information about synfuels on oil pricing. Benefits from both of these sources could be substantial, yet they are difficult to assess quantitatively.

Improved Information for Better Decisions

There remain substantial uncertainties regarding production costs, environmental impacts, and other issues for the major U.S. energy option represented by synthetic fuels. Initial production experience should clarify our information on key factors and provide an improved information basis for both future private investment and public policy decisions.

The benefits of improved energy policy and investment decisions are magnified by the fact that most substitutes for imported oil are relatively capital intensive. Oil is a convenient fuel that often can be displaced only by substituting capital (or in some cases operating and maintenance) expenditures for oil burning in meeting end-use energy needs. Since the variable costs of many production and consumption alternatives to oil are often low relative to the fixed costs, investment decisions often involve long-term commitment of substantial amounts of capital. Moreover, some energy production and consumption investments exhibit substantial economies of scale. Investments often must be made in large, discrete units rather than in small increments, and, once an investment is made, it generally represents a sunk cost, which cannot be converted economically to other uses. As a result, poor decisions and investments can lead to large inefficiencies and opportunity costs.

Major uncertainties surround society's many energy options. These uncertainties complicate and inhibit both private investment planning decisions and public policy. Improved information will reduce some uncertainties, allowing the private sector to better assess alternatives. With improved information, investment decisions can be both more cost-effective and expeditious. Over the long run, society's energy costs should be lowered. Private investment decisions also may be facilitated as improved information reduces the uncertainties and associated risk premiums that may inhibit investments in some energy alternatives.

In addition to providing an improved basis for private sector investment decisions, better information can foster improved national energy policies and encourage cost-effective investment of public resources to address our energy problems. Improved information can assist in ranking public policy options and focusing timely attention on critical issues. For example, clarification of the economic role for synfuels has implications for the appropriate direction for coal leasing policy. Also, better information on the relative cost and market role for important energy options can improve allocation of government budget resources among various research and development programs. If synfuels are a moderate-cost option, synfuels R&D investments become more likely to pay off. However, if synfuels are an $80 per barrel energy option, the nation is likely to be better off focusing limited R&D resources on other lower cost energy options. Better understanding of relative costs should clarify appropriate policy and program priorities.

The economic value of an improved information basis for private investment planning and public policy is difficult to quantify. A simple example calculation of possible benefits can be based on the cost savings obtained by reducing the average cost of the domestic alternatives used to displace one million barrels per day of oil imports by $1 per barrel. Net present value benefits resulting from this sample calculation are $4.5 billion, assuming a twenty-year investment life and 5 percent real discount rate.[4]

The point of this simple calculation is to illustrate the potentially substantial benefits to the nation of using relatively cost-effective energy options. There is no reason to expect that any particular private sector or government activity will yield such benefits. Indeed, the benefits of cost-efficient investments are produced continually by market processes and sound management practices. However, special attention to improving information on our energy options may be particularly important, given the large size of many investments, the substantial uncertainties regarding relative cost and performance, and the demonstrated potential for rapid rates of change in energy markets.

Practical Implications. What are the implications of these potential information benefits for an initial synfuel production program? First, because information can be valuable, it can be useful to acquire experience with major energy options that, while perhaps not yet economic, are likely to become important as market prices change. Second, for synthetic fuels in particular, experience from the first few commercial plants will reduce most of the uncertainty regarding the basic cost level for production of coal-based synthetic fuels and provide basic information on the environmental and socioeconomic effects of synthetic fuel production. It is important to note that the information outputs considered in this category are the very general observations that put the role of synthetic fuels in our energy future into perspective. (Detailed information about particular technologies is included in the category of the learning benefits for surge deployment discussed later.) These basic information benefits can be obtained largely from the first few initial production projects. A synfuel program designed primarily to obtain these general information benefits does not need to be very large.

Potential Oil Pricing Impacts of Information from Synthetic Fuel Production

Synthetic fuels are ready substitutes for petroleum products. Also, enormous hydrocarbon resource bases, such as coal and oil shale, could be applied to synfuels production. These two facts combine to suggest that the costs of synfuels could place a limit on the price of oil.

This discussion briefly considers some of the potential impacts on oil pricing that might result from the cost information derived from initial synthetic fuels production experience. The focus here is on the effect of this *information* on oil pricing. The physical role of synthetic fuel production—whereby production volumes reduce oil import demand and (depending on the production responses of oil producers) thereby reduce prices in the world oil market—is a separate effect considered earlier and included as a component of the oil import premium.

Economic theory supports the perception that oil prices and synthetic fuel production costs (or the costs for any other major source or substitute for petroleum) ought somehow to be linked (Hotelling 1931). Yet the linkages between expectations for the costs of synfuel production and oil pricing are not as strong or as simple as often perceived. Various economic paradigms for oil pricing suggest that the expectations of the owners of oil resources for the cost of major alternatives to petroleum should have some effect on rational oil pricing decisions (Hotelling 1931; Salant 1974; Weinstein and Zeckhauser 1975; Hoffman and Kennedy 1981). Recent OPEC long-range strategy documents (OPEC 1980) have indicated that price trajectories could be referenced to the expected cost of alternatives. For purposes of this discussion, the correctness of any particular model for rational oil pricing is not as important as the implications of potential oil pricing effects for the design of initial synthetic fuel production efforts.

The expected future cost of synfuels could affect oil pricing decisions even before the synfuels are produced. Expectations can be affected by changing information. Consequently, there are grounds to reason that the cost information from experience with synthetic fuels production could have some influence on the complex assessments that contribute to oil pricing and production decisions. This discussion outlines possible effects on oil pricing of information on synfuel production costs. A simple model is adequate for illustration.

Straightforward reasoning suggests that, if synthetic fuels can be produced in large volumes at constant costs, then synfuel production costs place an effective ceiling on oil prices. The level of synthetic fuel production costs acquires economic implications by defining a long-run limit to energy price levels. However, a major limitation of this simple model is that it is not clear how information about a ceiling level affects pricing and production decisions before that ceiling level is reached. With such a simple ceiling or "backstop" model, only actual production volumes at the ceiling price level affect oil prices. Mere *information* about the ceiling level would not have any clear effect on near-term oil price trajectories.

The economic theory of depletable natural resources provides the basis for more sophisticated models of rational economic oil pricing where there is a linkage between information about future cost levels and the economically "optimal" near-term price trajectory. Generally, these models apply economic reasoning by the owners of a depletable resource (such as oil) to determine pricing and production decisions that maximize some economic objective function. Oil in the ground is viewed as an economic asset for which profit-maximizing pricing and production decisions are related to the relative value of crude oil production now compared to value of production at some point in the future. The value of future production is related, among other things, to the future production cost for major alternatives to oil such as synthetic fuels. An economic linkage exists between producer expectations for the future value of their depletable assets and present production and pricing decisions. Consequently, information that affects producer expectations about the future value of their oil holdings could have immediate impact on oil production and pricing.

Although this theory demonstrates some linkage between information about the cost of oil alternatives and rational pricing decisions, numerous complex relationships and uncertainties make it impossible to definitively predict the direction, let alone the magnitude, of possible oil pricing changes in response to any particular information outcome.

Several factors limit our ability to predict the effect of information on oil pricing. First, oil pricing and production decisions may be determined by near-term political or physical factors rather than by long-run economic maximizing. Some combination of both may operate, but it is impossible to know the mixture. Furthermore, even under a pure economic approach, it is difficult to know the values for some key parameters (such as the preference for near-term spending as compared to saving for future generations) or how such parameters might change over time. Even more limiting, however, is the fact that, even if we know the economic model and key parameter values, the primary impact of information is to change existing expectations. Since existing expectations cannot be readily known, it becomes difficult to predict either the direction or the magnitude of the changes in oil pricing that might occur from new information.

These complications make it difficult to quantify potential oil pricing impacts from initial synthetic fuel production experience. We cannot even be sure whether various information outcomes will yield favorable or unfavorable changes in oil prices. Nevertheless, it may be unwise to completely ignore the potential interactions between information and oil pricing.

For example, synthetic fuels are commonly viewed as a "backstop" energy supply source that can be produced in infinite quantities at constant production cost.[5] Experiments with one model for oil pricing suggest that changes in expectations for backstop cost levels can lead to oil price trajectory changes on the order of hundreds of billions of dollars net present value.[6] Clearly, costs or benefits of changes in oil price trajectories potentially resulting from information effects are enormous. They are so large, in fact, that potential information effects have expected values that remain large relative to other sources of costs and benefits even if (1) one does not think it very likely that oil pricing follows some rationale economic model or (2) one does not think it likely that information from initial synfuels production will change the expectations underlying the decisions of oil pricers.[7]

Implications for Initial Synfuels Production Programs. Despite the difficulties of assessing the direction or magnitude of potential oil pricing effects stemming from the information acquired via initial synfuel production experience, a few lessons can be drawn for the design of initial production efforts. Biased information outcomes could have significant costs or benefits if expectations that figure in oil pricing are changed by the information. A synthetic fuel production effort that resulted in unnecessarily high costs for initial production plants (due to cost increases for inputs used to produce synfuels, construction inefficiencies, bottlenecks, etc.) could shift oil pricers expectations upward and lead to higher oil prices. For example, an initial production program that incorrectly increased expectations for synfuel cost from $40 to $60 per barrel could deter private domestic investment in synfuels as well as encourage more aggressive pricing by owners of oil resources.

Implementation of initial synfuel production efforts should give careful attention to the quality and management of the information outputs from the program. For example, if subsequent synfuel plants are expected to be less costly than initial synfuel facilities, it may be desirable to make the information on how and why that is the case readily available. It may also be desirable to control, structure, and interpret information from the program in a manner favoring U.S. interests. Although detailed assessment is beyond the scope of this study, design and implementation of initial synthetic fuel production efforts should seek to produce and manage the information outputs to yield information that is at a minimum unbiased, or, alternatively, perhaps even interpreted to favor U.S. interests. In any event, both the potential benefits and costs of information effects on oil pricing warrant further careful consideration.

SURGE DEPLOYMENT BENEFITS

A Key Source of Strategic Insurance Benefits

Recent experience has demonstrated the potential for rapid change that is inherent in the economic and political structure of the oil market. OPEC has displayed market power and cartel discipline sufficient to manage the pricing and production decisions that can result in rapid price changes. Since, in the short run, demand for oil cannot be readily reduced, small shortages in supply can lead to large price increases. The Persian Gulf area produces over half the petroleum traded in world markets. The well-known political volatility in that region has both historic and modern roots. The potential for domestic, regional, or international developments in the Persian Gulf to substantially change the market price or security cost of oil has been demonstrated several times in recent years.

Such rapid change stands in marked contrast to the pattern of market evolution predicted by economic reasoning applied to well-functioning competitive markets. Such reasoning would predict evolutionary change in oil markets with new, more costly sources gradually supplanting less costly supplies as they were depleted. Changes would be continuous and incremental, with steadily increasing oil prices resulting in gradual experimentation and adoption of new supply sources or demand adjustments. However, the potential for rapid or discontinuous rates of change in oil markets presents special challenges to the economic adjustment process.

Rapid changes in the market price or economic security costs of oil could quickly change the outlook for technologies and energy options that were previously unattractive relative to oil. All-out, or or "surge," efforts could be made to deploy energy options that suddenly become economic.

During such a "surge deployment effort" (SDE), growth of a synthetic fuels industry would not reflect a normal pattern of incremental evolution in production capacity and technology. Rather, many synthetic fuel plants would be built simultaneously. Prior experience with synthetic fuel production can enhance society's capability to undertake these surge deployment efforts yielding the three categories of surge deployment benefits: *learning, enhanced deployability*, and an improved *growth basis*.

Surge deployment benefits are similar to insurance payoffs because they are primarily realized during the special eventuality represented by an "emergency" deployment. The improved capabilities with the synthetic fuel option—capabilities obtained via initial production experience—are primarily important if a "crash" deployment effort becomes warranted.

Rapid change poses special issues for the growth of new industries. With gradual change in energy markets or growth in synfuel production capacity, strategic investments to realize learning are less important, since each relatively small increment to production capacity benefits from the accumulated knowledge of preceding plants. Gradual change also reduces the concerns for developing industry infrastructure, since the production increments required in any period may not be large relative to existing production levels. Finally, steady industry growth reduces the concerns for an improved growth basis in order to reduce the socioeconomic strains and costs often imposed by rapid growth. In gradually evolving markets, normal private activities in response to market opportunities should lead to improving technologies, expanding production capabilities, and increasing levels of economic activity and infrastructure to support industry expansion.

However, several aspects of energy markets in general and synthetic fuels in particular may make special attention and strategic investments for synfuels warranted. First, the potential rates of change in energy markets are demonstratably quite rapid relative to the minimum five-year time period required to obtain even basic experience with synthetic fuels as well as with many other major energy options. This magnifies the value of the information, learning, and infrastructure development realized via initial production experience.

Second, the fixed capital intensity of many alternatives to oil further amplifies the value of information and experience to respond to rapid change. Once built, a synthetic fuel plant cannot readily be modified to a better design, nor can the billions of dollars of capital required for a single synfuels plant be redirected toward more cost-effective oil substitutes, such as conservation efforts or heavy oil production.

Finally, the societal costs or externalities of rapid growth are particularly problematic for synthetic fuels.[8] Some resource bases for synthetic fuel production are located in sparsely populated regions. Intensive efforts to realize surge capacity expansion are likely to incur the special socioeconomic costs of rapid growth in locales with relatively low levels of pre-development population, infrastructure, and economic activity.

These effects—the potential for rapid market change relative to learning times, capital intensity, and the externalities of rapid growth—combine to magnify the strategic insurance value of improved national capabilities and posture to respond to our uncertain energy future.

Why Aren't Market Incentives Alone Adequate? Many of the benefits of enhanced national capability to respond to rapid change in energy markets can yield profit opportunities for private firms. For several reasons, however, private market incentives alone may not be adequate.

First, effective markets for many of the benefits of improved surge deployment capabilities do not exist. The private firm whose initial synthetic fuels plant provides cost-reducing experience to architect/engineering firms and component suppliers cannot capture as revenues many of the benefits stemming from his pioneering project. Also, some of the benefits of initial production experience are largely public goods. For example, much of the information stemming from pioneering synthetic fuel experience cannot be readily privately held or marketed for profit.

Second, surge deployment benefits are insurance-like payoffs realized primarily in extreme and (hopefully) unlikely scenarios. The exposure from a single project exceeds the net worth of all but the largest corporations. Risk-averse investors may understandably hesitate to undertake multi-billion dollar investments that payoff primarily as insurance against adverse scenarios for energy markets. The nation may not wish to be as risk-averse as private investors. Moreover, the nation as a whole can capture the full insurance payoff, rather than only part.

The value of such insurance investments depends on the probability for the unfortunate scenario. Intelligence systems and foreign and domestic policy options may allow government to have both better information and more influence than private firms on the probability of scenarios that might lead to surge deployment efforts;[9]

Moreover, even if a surge deployment event were to occur, private firms may not expect that they can realize the full profitability of investments made to hedge against the event. Adverse developments in the market price or security costs of oil might invoke traditional patriotic proscriptions against "profiteering" during a national crisis. "Windfall" taxes might limit the private profitability, but not the value to the nation, of insurance investments in pioneering production projects. The mere possibility of such reductions in profit potential would reduce private sector interest in proceeding with initial production projects (Schelling 1979: 41-45).

The incentives of the private market place alone, therefore, may not lead to appropriate levels and forms for the experience and capabilities that could yield substantial benefits to society should an intensive expansion of synthetic fuel production capacity become necessary. Some special government efforts and incentives beyond the marketplace may be justified. Nevertheless, since existing capabilities and normal unsubsidized private investments in initial production projects will provide some capabilities to support surge deployment, the *net* benefits of any government program should be assessed relative to a base case reflecting the private investment, activities, and capabilities that would be expected without any special government efforts.

Specific Rationale and Guidance for Synfuels. The potential for rapid change in energy markets creates the possibility for surge deployment of energy options besides synthetic fuels. However, the benefits of improved capabilities for surge deployment scenarios provide a rationale for special attention to synthetic fuels because of the long lead-times required to acquire basic experience and improved capabilities. Strategic investments in surge deployment capabilities are less important for energy options for which lead-times for learning are short. The large size of individual synfuels facilities and the difficulties of modifying them if design mistakes are made further magnify the value of improved capabilities for surge deployment scenarios.

The production benefits of the oil import premium and the information benefits provide no strong rationale for special attention to the synfuels option. The surge deployment category of benefits, however, is relatively specific for large, capital-intensive, long learning time energy options such as synthetic fuels. The positive externalities of initial production experience provide an economically-grounded basis for special investments beyond the marketplace. Moreover, the insurance benefits of improved surge deployment capabilities suggest important issues and guidance for appropriate program purposes and design. Accordingly, surge deployment benefits are a central concept in this study. The remainder of this section describes and quantifies this important category of benefits further. Learning and deployability benefits receive primary attention. Parametric quantitative estimates are presented to assist later comparisons of the relative size of benefits and costs.

Learning Benefits for Surge Deployment

Presently, U.S. industry has only very limited experience with the complex activity of siting, designing, constructing, and operating the facilities for synthetic fuel production.[10] Consequently, the value of learning from initial production experience could be magnified should a surge deployment effort become necessary.

This section describes the learning benefits which are likely to be obtained from an initial synthetic fuels production experience. While this chapter emphasizes the definition and parametric quantification of learning effects, their implications for the design of initial synthetic fuel production programs are the focus of Chapter 9.

Initial synthetic fuels production experience will yield two important types of learning. First, information about previous projects and technology performance can help improve the selection of *which* technologies should be deployed. This "selection learning" may lower the average cost of the technology mix utilized.

Second, experience with synthetic fuels technology in general as well as a particular technology or process design commonly suggests both technology design and project execution improvements. When such "technology improvement learning" is avilable prior to a surge deployment effort, those plants built simultaneously as part of such an intensive deployment effort can incorporate the technology improvements. The resulting synfuel production capacity will incorporate "better plants." Costs will be lower.

This section describes each of these sources of learning benefits— selection learning and technology improvement learning. Parametric quantitative estimates of the magnitude of benefits are also presented in order to assist cost-benefit comparisons.

Selection Learning. Commercial synthetic fuels production could utilize a number of different technology and resource base combinations. Little real experience exists, however, upon which to differentiate cost and performance among the many alternative technologies that could be used for the multiple plants that would be built simultaneously as a part of a surge deployment effort. Initial commercial production experience should provide information that will allow improved technology selection.

The potential value of this technology selection information can be illustrated by a simple example using two alternative technologies, Technology A and Technology B. For simplicity, assume these technologies produce identical products and have similar environmental effects. Discrimination among the technologies should be based on cost alone. If pilot plant and paper engineering studies are accurate to only plus or minus 20 percent, then there would be no reason to select one technology over the other if these studies showed less than a 20 percent difference. Approximately equal use of each technology would be expected to be made in a surge deployment effort.

However, if commercial projects utilizing the two technologies reveal a significant cost difference, then a surge deployment effort would emphasize the lower cost process. If, as an example, a significant cost difference of 10 percent was observed, then a deployment based on an equal mix of Technology A and Technology B would be expected to have an average cost at least 5 percent greater than the average cost for deployment of a technology mix composed entirely of process A. Thus, improved project selection can reduce the expected average cost for the mix of plants constructed as part of a surge deployment effort.

The benefits of reducing the average production costs for the surge deployment technology mix can be substantial. A $1 per barrel cost saving for a nominal 50,000 barrel per day plant accumulates over the thirty-year life of the facility to a discounted present value (5 percent discount rate) of $0.25 billion in the year of initial operation. In the above example, the 1980 present value of the 5 percent reduction in average costs for the technology mix depends on several factors, including the synfuel cost level, the date at which surge deployment plants begin production, the production capacity of the synfuel plants built as a part of a surge deployment effort, and the discount rate. Table 2-1 presents illustrative parametric estimates of the benefits of a 5 percent reduction in the cost of the technology mix for a group of surge deployment plants beginning operation in the mid-1990s. Estimates in the third column display savings (net present value) for each nominal plant. Figures in the far right column assume a surge deployment effort of approximately one million barrels per day or twenty plants.

This example of a 5 percent average cost reduction is a substantial simplification of a complex situation. First, the example assumes that the initial commercial experience will show a significant cost difference between the two processes. Significant cost differences may not necessarily be observed. Also, it may be difficult to discriminate between cost outcomes inherent to the technologies and cost outcomes more closely related to the efficiency (or lack thereof) for project execution. Where well-grounded discrimination among costs is not possible, the selection of the surge deployment technology mix may not be substantially altered. In the extreme case, initial production experience would yield no benefits from selection learn-

Table 2-1. Example Benefits from Selection Learning (assuming average surge deployment costs reduced by 5 percent).

Synfuel Cost Level	Per Barrel Savings of 5% On Average	Net Present Value of Savings For Each Plant[a]	Total Benefits for 20 Surge Deployment Plants[b]
(1980 $/bbl)	(1980 $/bbl)	(Billions of 1980$)	(Billions of 1980$)
30	1.50	.19	3.7
40	2.00	.24	5.0
50	2.50	.31	6.2
60	3.00	.37	7.4
70	3.50	.43	8.6

[a]Assumes 5 percent discount rate for plant operating twenty five years, beginning operation in 1995. Each plant produces 50,000 bbl/day.

[b]Assumes surge deployment effort of twenty plants or 1 MMBD. Far right column is simply twenty times the column to its immediate left.

ing since the technology mix after initial production experience would be the same as the technology mix expected earlier.

The quantitative assessment of the expected benefits from selection learning is a complicated process requiring detailed specification of a number of factors, including the number of technology alternatives and our ability to distinguish between initial production cost outcomes due to inherent technology performance and those due to project execution effects. Nevertheless, the value of selection learning as a potentially significant source of benefits from initial synfuel production is clearly suggested by the example estimates in Table 2-1. Benefits from selection learning could plausibly reach totals approaching billions of dollars (net present value). Further assessment of selection learning effects is warranted. Chapter 9 considers these effects, with emphasis on the implications of selection learning for improved design and implementation of initial synthetic fuel production efforts.

Learning Curve Technology Improvements. Selection learning may lower costs by clarifying *which* technologies should be used in a surge deployment effort. "Learning curve technology improvements" lower expected costs for production capacity built during a surge deployment effort by providing better understanding of how to

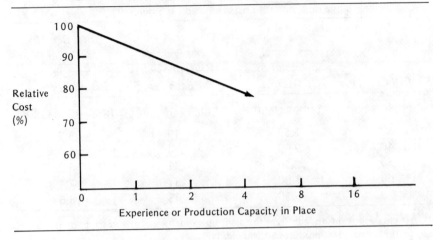

Figure 2-1. Stylized Hypothetical Learning Curve

proceed with a particular technology and project. *Better versions* of whatever technologies are used can be built. For quantitative assessments, this study uses a stylized learning curve as a summary representation of the cost reductions that become possible as experience leads to increased understanding, higher efficiency, and reduced per unit costs.[11]

Stylized learning curves, such as that shown in Figure 2-1, summarize a relationship between changes in production costs and the amount of production experience. The sloped line in Figure 2-1 illustrates the per unit cost reductions achieved with each doubling of the production capacity utilizing a technology. Total production capacity often is used as a quantitative representation of the amount of experience. Under gradual market evolution, one would expect to move steadily down the curve pictured in Figure 2-1 as each capacity increment yields the expanded information, experience, and understanding that foster cost-reducing improvements applied to later production increments.

During a surge deployment effort, however, large capacity increments are made simultaneously. Plants built as part of an intensive deployment effort must be based on the existing knowledge about technologies. Opportunities for incremental, sequential learning are limited. Figure 2-2(a) illustrates a learning curve/capacity relationship for a surge deployment effort.[12] Experience with synthetic fuels prior to a surge deployment effort "buys" progress down the learning curve. Consequently, surge deployment capacity expansion

Figure 2-2(a). Learning Curve/Capacity Relationship for Surge Deployment Effort

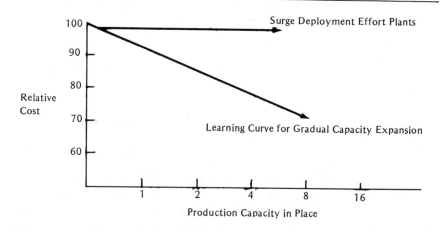

Figure 2-2(b). Surge Deployment Cost Levels Reduced Via Learning

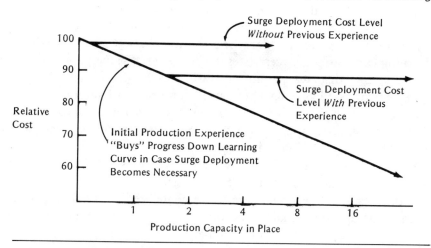

can be accomplished at a lower cost level, as illustrated by Figure 2-2(b).

Estimating the Value of Technology Improvement Learning. The quantitative value of potential surge deployment cost savings depends on three major factors. First and most important is the size of the cost reductions from the technology improvements realized via practical production experience. The slope of the stylized learning curve quantitatively summarizes this relationship between experience and per barrel cost reductions. Second, the present value of cost savings for a single surge deployment plant depends on when a surge deployment occurs and on the discount rate being applied for quantitative analysis. (This study assumes a 5 percent discount rate.) Finally, total technology improvement benefits depend on the size of the surge deployment effort. Unfortunately, all three of these major factors are uncertain. Accordingly, this section develops parametric estimates that illustrate a range of possible magnitudes for technology improvement learning benefits associated with a surge deployment effort.

Learning curves summarize the linkage between production experience and declining per unit production cost. For capital intensive chemical process technologies, learning curve effects often are summarized as a percentage reduction in the per unit capital cost (not total cost) for each doubling of experience with a technology (Merrow 1978: 43-55). Thus, the expected dollar magnitude of learning curve savings depends on the capital cost level for synfuels, the amount of previous experience, and the percentage cost reduction as a function of experience. Uncertainty about the magnitude of learning curve effects necessitates a wide range of values.

Table 2-2 displays parametric estimates for learning curve cost savings (expressed as dollars per barrel) for a range of synfuel production costs and three different assumed slopes for learning curves. Table calculations apply learning effects to per unit capital costs only. Synfuel capital cost is calculated as assumed total production costs less $14 per barrel for coal feedstock costs and $5 per barrel for operating and maintenance costs. Learning is primarily a sequential process. Most technology improvements are identified only after plant start-up and initial operation. Thus, a unit of experience correlates most closely with the number of previous sequential groups or waves of plants. The range of learning curve slopes, from 2 to 15 percent, encompasses many informed expectations for learning rates with capital-intensive energy process technologies.[13]

Table 2-2 shows that the amount of cost savings is a strong function of the learning factor and synfuel cost level. Moreover, the relationship between cost-reduction and experience is such that doubling the amount of sequential experience would nearly double the per barrel amount of cost savings expected for a surge deployment effort. Representative values for learning curve cost reductions appear to be about $1 to $5 dollars per barrel.

Table 2-2. Potential Per Barrel Amounts of Technology Improvement Cost Savings (cost reduction in $/bbl for various parameter values).

Synfuel Cost Level ($/bbl total cost)		Learning Factor (percentage reduction for each unit of experience)		
		2%	8%	15%
30	(11)[a]	0.22	0.88	1.65
40	(21)	0.42	1.68	3.15
50	(31)	0.62	2.48	4.65
60	(41)	0.82	3.28	6.15
70	(51)	1.02	4.08	7.65

[a]Numbers in parentheses indicate capital cost assuming coal feedstock and O&M costs totaling $19 per barrel ($14/bbl for feedstock assuming coal at $1.50/mm Btu and 67 percent conversion efficiency). For background, see Chapter 3.

These dollar per barrel cost savings accumulate over the life of a surge deployment plant to yield total expected benefits per representative plant. The net present value in 1980 of the savings for each surge deployment plant depends on when a surge deployment effort occurs. Table 2-3 presents parametric estimates of the total technology improvement learning benefits for a representative surge deployment effort plant beginning operation in various years. Total per plant benefits increase in direct proportion to the dollar per barrel savings estimated in Table 2-2. Thus, for example, the figures for a $5 per barrel cost savings in the right hand column are simply five times the figures in the column showing per plant benefits of each dollar of learning curve cost reductions. Table 2-3 illustrates the

Table 2-3. Benefits Per Plant from Technology Improvement Learning (billions of 1980$ net present value at 5 percent discount rate).

Timing of Surge Deployment Effort		Size of Per Barrel Cost Saving	
Year Initiated	Year Plants Begin Production	For Each Dollar	For $5 Per Barrel Savings
Mid-1980s	1990	.163	0.82
Late-1980s	1995	.127	0.64
Mid-1990s	2000	.099	0.50
Late-1990s	2005	.078	0.39
Mid-2000s	2010	.061	0.31

sensitivity of the net present value of learning benefits to when a surge deployment occurs. However, for plausible dollar per barrel cost reductions, the 1980 net present value of learning benefits could approach $0.5 billion for each surge deployment plant.

Note that synfuel plants coming on line before the early 1990s will have limited opportunity to benefit from initial production experience, since they must be substantially designed and under construction before practical operating experience is available from projects begun in the early 1980s.

The total benefits from technology improvement learning depend on the size of the surge deployment effort. The per plant estimates of Table 2-3 are simply multiplied by the number of surge deployment plants to estimate total benefits. Alternative assumptions about when a surge deployment effort occurs and the size of the per barrel cost savings can be related readily to the estimates of Table 2-3. For example, for a surge deployment effort initiated in the late 1980s and comprised of thirty "nominal" plants,[14] total benefits would equal thirty times about $0.6 billion (from Table 2-3), if we assume initial production experience yielded technology improvement learning cost savings of about $5 per barrel. Total benefits for this example would equal about $18 billion net present value.

Such calculations illustrate the substantial benefits that could be realized from the capability to undertake surge or emergency expansion of synthetic fuel production capacity at lower costs. Although benefits vary substantially with assumptions for key uncertain parameters or scenarios, the total social benefits from "better" surge deployment plants could plausibly total billions or tens of billions of dollars (net present value).

Total Learning Benefits for Surge Deployment. The benefits from selection learning and technology improvement for surge deployment should be largely additive. Selection learning lowers expected average cost of the technology mix and yields benefits even if no technology improvement learning occurs. Technology improvement learning lowers the expected costs for those technologies that are utilized, even if no selection learning is realized. Accordingly, total surge deployment learning benefits are approximately the sum of savings from the two sources.[15]

These parametric estimates for two important sources of learning benefits suggest that *if* a substantial surge deployment effort occurred, total learning benefits could plausibly reach net present values on the order of tens of billions of dollars. We shall see in Chapter 3 that such substantial benefits, in the event of a surge deployment effort, could offset and exceed the potential subsidy costs for initial production projects. Some subsidy costs may be warranted as a strategic investment to foster the initial production experience which leads to the improved information and capabilities that are the source of surge deployment learning benefits.

However, "if" is an important word in the cost-benefit calculus. Learning benefits are like insurance payoffs; they accrue only if a surge deployment or insurance event occurs. If no surge deployment occurs, no learning benefits offset the subsidy costs for initial production. Moreover, the size of learning benefits depends on a number of uncertain parameter values, such as synfuel costs and learning rates, as well as on the size and timing of the surge deployment scenario.

Although their magnitude is uncertain, potential surge deployment learning benefits could be large enough to significantly counterbalance any subsidy costs for initial synfuel production efforts. They are an important element of a framework for the benefits and costs of initial synthetic fuel production projects.

Enhanced Deployment Capability

Overview. The United States has an enormous and reasonably well-dispersed coal resource base. Theoretically, this resource base could support large volumes of synthetic fuel production; however, at least initially, there are significant constraints on the number of production facilities that can be simultaneously built to convert coal resources into synfuel substitutes for petroleum. Put simply, large industries cannot be created overnight.

Rapid increases in the market price or security costs of oil imports could motivate an effort to simultaneously deploy large numbers of synfuel plants. A large industry would be sought without the usual time for evolutionary infrastructure development. Deployment constraints would limit the size of the production increments that could be realized during such an all out surge deployment effort.

Experience with synthetic fuel production should build infrastructure and expand deployment capabilities. "More plants" could be built during an all-out deployment effort. Each additional plant would provide substantial social benefits. Moreover, expanded industry infrastructure could reduce added costs from the factor cost inflation—due to bottlenecks, price increases, and project execution inefficiencies—that intensifies as the level of deployment begins straining existing capabilities.

Both types of benefits from enhanced deployment capability—"more plants" possible and reduced factor cost inflation—could yield substantial savings to the nation should a surge deployment effort become necessary. Thus, *enhanced deployability* is another important category of insurance benefits from initial synthetic fuel production efforts. This section describes enhanced deployability benefits further and develops some quantitative estimates of their possible size.

The synfuel production volumes realized during a surge deployment effort can be increased in two ways. First, additional parallel plants may be built by relaxing constraints limiting the number of simultaneous synfuel projects that can be undertaken. Second, reductions in project completion time can increase the cumulative production achieved in a given time period.

This study limits its attention to the first approach toward enhanced deployment capability. It focuses on increasing the volume of production by increasing the number of parallel or simultaneous plants that can be constructed. Reducing the time required to procede from project initiation to plant start-up does not necessarily increase the number of simultaneous projects possible or the production volumes reached at the end of the initial phase of a surge deployment effort.

Reduction of a time required for a synfuels project has been the focus of procedural reforms—such as the Energy Mobilization Board concept considered in the 96th Congress—as well as site banking, pre-design, and other proposals. Such efforts may be a cost-effective way to increase the benefits from synfuels in both surge deployment efforts or more evolutionary scenarios. Procedural reforms and expediting efforts deserve attention and analysis, but they have been excluded from the scope of this study, which considers only the benefits of "more plants" (rather than "faster plants") during a surge deployment effort.

Constraints on the number of simultaneous synfuel plants possible could arise from many sources. These are characterized in Chapter

10. It is widely recognized, however, that bottlenecks and constraints for key service or hardware inputs during a surge deployment effort could lead to delays, inefficiencies, price inflation, and other problems. Such developments would effectively limit the maximum number of projects that could be simultaneously constructed as well as increase the costs for all the plants built in a surge deployment effort.[16]

The point at which existing infrastructure constraints would limit surge deployment efforts remains uncertain, but initial synfuel production experience should help build the industry infrastructure and expand the supplies of the inputs required to deploy synfuel plants.[17] Constraints should be relaxed and deployment capability increased over existing levels.

This study does not make detailed assessments of particular deployment scenarios and possible constraints. Chapter 10 examines the implications of attention to deployment capability for design of an initial synthetic fuels production program. This section focuses on developing parametric estimates of the possible magnitude of the benefits from enhanced deployment capability should an all-out surge deployment effort become warranted.

Estimating The Benefits From "More Plants": An Overview. Figure 2-3 illustrates the general concept for enhanced deployability benefits. Prior experience with synthetic fuels production should allow larger production increments (Curve A) in a surge deployment effort than would otherwise be possible (Curve B).

Estimates of the social benefits of enhanced deployment capability require specification of two major factors. First, the volume of additional production made possible by enhanced deployment capability must be estimated. This is the shaded area between the two production growth curves in Figure 2-3. Second, the benefits of the additional production volume must be estimated. The benefits to the nation of each additional barrel equal the difference between synthetic fuel production costs and the social cost of a barrel of imported oil. This social cost of oil imports comprises the market price plus an oil import premium.

After specifying these two quantities, total enhanced deployability benefits may be estimated with a straightforward calculation of quantity times cost or value. Since the social cost of oil may

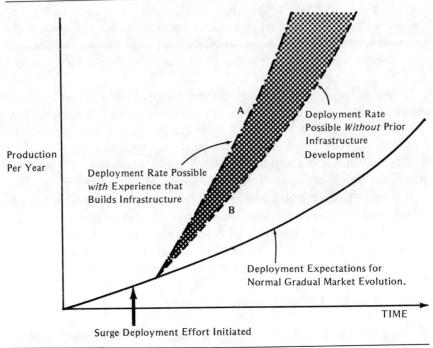

Figure 2-3. Illustration of Concept for Enhanced Deployability Benefits

change over time, year-by-year calculations using assumed oil price trajectories are necessary. The net present value of deployability benefits equals the summation with appropriate discounting of the year-by-year calculation of quantity times value.

Both of the major elements that underlie benefit estimates—quantity differences and cost differentials—are substantially uncertain. Benefit estimates thus depend on particular parameter and scenario assumptions. Each of these two major elements that determine the total benefits of enhanced deployability is examined in turn below.

Quantity differences are examined first. A simple algebraic model of the linkage between practical production experience and expanding deployment capability is used to define ranges for the number of additional plants that might be possible in a surge deployment effort.

The value of each additional plant possible—the second major element—depends on the cost differential between synfuels and oil over the lifetime of each surge deployment plant. The net present value of these cost differences are estimated for some stylized surge deployment scenarios. This requires specification of some example long-run oil price trajectories as well as some stylized price jump events above those long-run smooth trajectories. The value of each synfuel plant also depends on the uncertain synfuel production cost. Tables are used to display the net present value of each synfuel plant for a range of synfuel production costs and oil price scenarios.

Given the many uncertain parameters that underlie estimates of both quantity differences and cost differences, illustrating the estimation of enhanced deployability benefits is a long process. The sections below present basic details in sufficient depth to allow interested readers to follow the logic and calculations. However, some readers may be less interested in the necessarily stepwise development of estimates for enhanced deployability benefits. Those readers may wish to skim or skip the next two sections and focus on the summary estimates for the benefits of enhanced deployment capability.

The conclusion of the benefit estimating process is straightforward. The ability to build an additional plant in a surge deployment scenario where the social costs of oil exceed synfuel production costs can result in benefits amounting to about 1 billion dollars (net present value) for each extra plant. The number of additional plants possible depends on the size of the initial production effort and the strength of the quantitative linkage between production experience and expanding future deployment capability. If, for example, ten additional surge deployment plants are made possible by an initial production program, total enhanced deployability benefits could reach tens of billions of dollars (net present value). Like surge deployment learning benefits, enhanced deployment capability is an important category of insurance benefits that, if a surge deployment effort occurs, can offset the subsidy costs required to promote initial synthetic fuel production projects.

How Many Additional Plants Are Possible? Initial synfuel production experience should increase the number of simultaneous plants possible in a surge deployment effort, but the extent of the increase is largely speculative at this point. Nevertheless, it is possible to describe some plausible attributes of the relationship between prior production experience and expanding deployment capability and thereby develop a basis for some rough quantitative reasoning.

Maximum deployment capability is composed of two elements. The first element is "basic deployment capability." Although we have no substantial production experience with synthetic fuels, we could deploy some plants now. Basic deployment capability reflects this innate capability of the U.S. economic infrastructure to undertake major petrochemical construction projects such as synthetic fuels plants. Because it is "basic," basic deployment capability is a constant that does not depend on the presence, absence, or extent of prior production experience. Without prior production experience, maximum surge deployment capability equals basic deployment capability.

The second element is the building of expanded infrastructure via practial production experience. Such infrastructure development should increase maximum surge deployment capability. The amount of increase depends on both the amount of prior production experience and the strength of the linkage between production experience and expanding future deployment capability. If the linkage is strong, small to modest initial production efforts provide large increases in maximum surge deployment capability. If the linkage is weak, larger initial production programs may be needed to significantly increase the infrastructure to support synfuels deployment.

This relationship between increasing deployment capability and production experience also has two elements. The first is the level of prior production experience. *Additional* deployment capability probably increases in direct proportion to the level of prior experience. However, the strength of the linkage between deployment capability and experience may exhibit a "critical mass" relationship. A large initial production effort might exercise (or even strain) the engineering industrial system required for synfuels deployment much more extensively than a small or modest program. Reflecting a chain reaction effect, the linkage between initial production experience and future deployment capabilities could be relatively stronger for higher levels of initial production activity.[18] The result of this plausible relationship is that larger initial production efforts increase deployment capability more than in direct proportion to their size. Expanding deployment capability may evidence increasing marginal returns to scale. (This stands in contrast to learning benefits, which are likely to exhibit decreasing marginal payoffs for larger initial production program size. The implications of these contrasting benefit-program size relationships are considered as part of the "How large?" question addressed in Part II of this study.)

A simple hypothetical model for the relationship between maximum deployment capability and production experience can reflect the qualitative relationships noted above. One simple model has the form:

Maximum Deployment Capability	=	Basic Deployment Capability	+	(Deployability Increase Factor)	×	(Amount of Previous Production)

or, in algebraic symbols, $MDC = BDC + (DIF) \times (APP)$. This formula can be used to develop some illustrative numbers for the number of additional plants possible from the enhanced deployment capability generated by initial production experience. Maximum deployment capacity (MDC), basic deployment capability (BDC), and amount of previous production (APP) can be measured as the number of simultaneous nominal synfuel plants. The deployability increase factor (DIF) is a simple linear factor that defines the number of additional plants possible for each previous plant. Thus, if DIF equals 1.0 and an initial production program were comprised of ten plants, ten additional plants would be possible in a subsequent surge deployment effort. The total number of plants possible in the surge deployment effort equals the additional plants *plus* the basic deployment capability.

Table 2-4 displays some illustrative calculations using this simple linear formula for three different initial production program sizes: five, ten, and twenty initial production projects. Reflecting an assumed critical mass effect, DIF values are assumed to be 0.5 for the five plant program, 1.0 for the ten plant effort, and 2.0 for the twenty plant program. Entries in Table 2-4 are expressed in millions of barrels per day assuming 50,000 barrels per day for each nominal plant. Entries in parentheses indicate the corresponding number of plants. Columns show additional production possible, total surge deployment production assuming a basic deployment capability of twenty plants (or 1.0 MMBD), and cumulative production level.

Table 2-4 illustrates the wide range of variation in deployment capability possible with different initial production program sizes and expectations for the deployability increase factor. The assumed value for basic deployment capability (BDC) affects cumulative production levels, but does not affect the number of additional plants possible. Estimates of enhanced deployability benefits should be based on the number of the additional plants possible. A large

Table 2-4. Example Synfuel Production Levels with Enhanced Deployability (quantities in millions of barrels per day oil equivalent).

Program	Additional Production from Enhanced Deployability	Total Surge Deployment Size (BDC = 1.0 MMBD)	Cumulative Production Including Initial Production Program
5 Plants with DIF = 0.5	.125 (2.5)	1.125 (22.5)	1.375 (27.5)
10 Plants with DIF = 1.0	.5 (10)	1.5 (30)	2.0 (40)
20 Plants with DIF = 2.0	2.0 (40)	3.0 (60)	4.0 (80)

Note: Entries in parentheses indicate the corresponding number of nominal synfuels plants.

number of plants may be possible without prior production experience if basic deployment capability is large.

The deployability increase factor (*DIF*) represents the quantitative linkage between initial production experience and expanding deployment capability. It has a strong influence on expected benefits. If *DIF* is doubled, the number of additional plants possible is doubled. Unfortunately, our present understanding of the phenomena which provide a basis for estimating the strength of the *DIF* factor is limited. Chapter 10, however, develops the view that engineering and mega-project management teams capable of undertaking a complex multibillion dollar synfuels project are a key limitation on synfuels deployability.

Under this view of deployability, if one initial synfuels project trains one more synthetics-capable architect, engineer, and constructor (AE&C) team, *DIF* values would be expected to be about 1.0. Each initial production plant might be plausibly expected to allow about one additional plant to be built should a surge deployment become necessary.

Total quantity differences will depend on the size of the initial production effort and on the value of the *DIF* linkage factor. Depending on particular program sizes and assumptions, the number of additional plants possible could range widely from only a few to twenty or more.

So far we have considered only the quantity of additional synfuel production possible via improved deployment capability. Total benefits from enhanced deployability depend also on the value of those additional production volumes as discussed below.

How Much Is Each Additional Plant Worth? If surge deployment is motivated by marked increases in market price or security costs of imported oil, each additional plant possible should yield some benefits. Estimates of the value of the incremental production quantities depend primarily on the effective cost differential between synfuels and oil. Both of these key parameters are uncertain. However, by assuming oil price scenarios and repeating calculations for a number of synfuel cost levels, parametric tabulations of the net benefits from each additional surge deployment plant can be developed. Such tables can illustrate plausible ranges for the benefits (net present value) for surge deployment plants deployed in response to various oil price jump or security event scenarios. This section develops these parametric estimates for one set of oil price trajectories and a range of synfuel cost levels.

Estimates of the net present value of each additional synfuel plant are based on a straightforward calculation that compares the cost of oil to synfuel production costs over the lifecycle of the synfuel plant's operation. Benefits or costs in any given year equal imported oil costs *less* synfuel production costs. This difference is multiplied by the volume of production in any given year. Lifecycle net present value is determined by adding the yearly costs or benefits with appropriate discount factors to reflect the time value of money.

Synfuel Production Costs. For cost-benefit estimates, synfuel production costs can be expressed on a constant or levelized basis. Synfuel costs should include the full cost to society of the real resources employed for synfuel production *plus* the imputed costs of any environmental or social externalities not reflected in the investment and production cost accounting. Parametric tabulations are developed by repeating calculations for a range of synfuel costs.

The Social Cost of Oil. Estimates for the second key determinant of the value of additional deployment capability—the oil price trajectory realized over a surge deployment plant's lifecycle—are more complicated.

The future trajectory for oil prices is highly uncertain. Even in a perfectly economic world devoid of political effects on oil pricing, the trajectory for oil prices would depend on numerous uncertain factors, such as resource endowments, economic growth, technological change and the costs for oil substitutes, and the time preferences of producers and consumers (discount rate). Political factors further complicate the underlying economic uncertainties.

The effective cost of oil imports is composed of two elements—the market price and the "premium," which includes implied security costs of reliance on oil from international markets. Sudden increases in either of these two elements could motivate a surge deployment of synthetic fuels. A sudden increase in the market price of oil will be called a "price jump event." A "security event" is a change in the imputed security costs for oil purchased on world market due to a fundamental change in long-run geopolitical assessments of the security liabilities associated with international oil trade. An example of a security event would be Soviet takeover of Saudi Arabia even if oil prices and production volume did not change. The value of a synfuel plant deployed in response to either a price jump event, a security event, or both simultaneously depends on the underlying long-run oil price trajectory. The value of an additional plant displayed in response to an oil price jump event will be estimated first.

Oil Price Jump Events. For purposes of cost-benefit analysis, long-run oil price trajectories are assumed to be smooth, gradually evolving paths that reflect the underlying geologic, technical, and economic determinants of oil prices. These factors are assumed not to be subject to a rapid change or political manipulation. Accordingly, the long-run path for oil prices reflects steady progress along a smooth trend. Given the fundamental uncertainties, parametric assessments should assume a range of long-run smooth trajectories.

This study uses a set of long-run price trajectories that have been applied by the U.S. Department of Energy (DOE) as a set of standard price cases for consistent program and policy assessment. These Policy and Fiscal Guidance price cases reflect various descriptions of the oil market over time and are designed to cover a reasonable range of alternative expectations about the long-run evolution of the oil market. Table 2-5 displays these 1980 DOE Policy and Fiscal Guidance price cases (DOE 1980b).

Stable markets would lead one to expect uncertain, but smooth, oil price trajectories such as the schedules in Table 2-5. The experience of the 1970s, however, demonstrated the potential for rapid, discontinuous changes in oil prices. Rapid oil price increases have occurred before and could occur in the future from supply reduc-

Table 2-5. Department of Energy Policy and Fiscal Guidance Price Trajectories (refiner acquisition cost of imported oil in 1980$ per barrel).

Case	Year					
	1978	1980	1985	1990	2000	2020
High prices	18	35	45	50	55	60
Medium prices	18	30	35	40	45	50
Low prices	18	25	25	30	35	40

tions due to: disruptions, such as the Iranian revolution or Iran-Iraq war; production decisions, such as those of Saudi Arabia; or many other events. Such major supply reductions can induce price increases because of the difficulty of reducing oil demand in the short-run (demand is inelastic). Also, long-run supply responses to higher oil prices often involve major capital investments requiring many years or perhaps decades to plan and implement. Oil supply cutbacks could lead to sudden price jumps. Increased oil costs could be maintained or might decay slowly until returning to a long-run trajectory reflecting some more fundamental underlying resource economics for the supply and demand of oil and its alternatives.

This notion of a long-run trend for oil prices along with possible deviations above a smooth trend is explicitly reflected in a recent long-run strategy document of the Organization of Petroleum Exporting Countries (OPEC). Long-run price trends were related to resource endowments, economic growth rates, rates of return on investment, and the costs of alternatives to oil. The possibility of shortages leading to price jumps above the long-run trend was noted. One option following a price jump event was to "freeze prices in real terms until the floor [i.e., the long-run trend] catches up with the new price level" (OPEC 1980: 10-12).

Price jump events and a return to a long-run trajectory could be described in any number of ways. However, quantitative analysis requires assumption of some stylized descriptions. This study represents price jump events as deviations above the long-run trajectories that last approximately ten to twenty years. The ten-to twenty-year period reflects the view that, over one or two decades, capital stock turnover and other energy supply and demand responses to changing price levels will cause oil prices to return gradually to some underly-

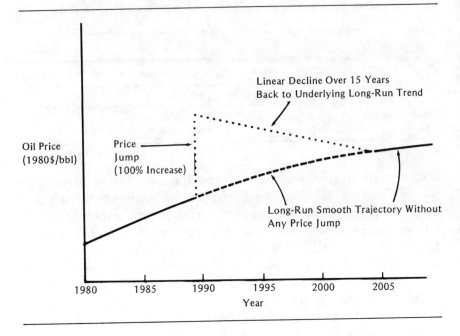

Figure 2-4. Illustration of Stylized Modeling for Price Jump Event

ing long-run trend.[19] Moreover, reasonably long-lasting (rather than temporary) price increases are needed to motivate investments in alternatives, such as synfuels, that take years to deploy.

Any number of stylized depictions of price jump events are plausible. However, for estimating purposes, a standard price jump will be described as a sudden doubling of oil prices followed by a linear decline over fifteen years back to the underlying long-run trajectory. Figure 2-4 illustrates.

Such rapid price increases could motivate a surge deployment effort. Production plants deployed in response to a price jump begin production some years after the price increase. Given the time required to site, design, and construct synfuel plants, a lag of approximately seven years is likely (Flour 1979: 31). For example, year-by-year calculation of benefits from synfuel plants deployed in response to price jump in 1985 would start as the surge deployment plants began production in approximately 1992.

To summarize, the net benefits of each additional surge deployment plant built in response to a price jump event depends on several parameters: the descriptions of long-run oil price scenario and price

Table 2-6. Net Present Value for Each Surge Deployment Plant.[a] DOE-PFG Price Trajectories for Price Jump Events in Year Indicated[b] (NPV of Savings or (Subsidy) in Billions of 1980$ at 5% Discount Rate)

Synfuel Cost Level (1980$/bbl)	Low Oil Prices			Medium Oil Prices			High Oil Prices		
	1987	1993	2000	1987	1993	2000	1987	1993	2000
20	2.7	2.3	1.6	4.4	3.5	2.4	5.9	4.7	3.1
30	1.3	1.2	0.9	2.9	2.4	1.7	4.5	3.6	2.4
40	(0.1)	0.2	0.3	1.5	1.4	1.0	3.1	2.6	1.8
50	(1.5)	(0.9)	(0.4)	0.1	0.3	0.4	1.7	1.5	1.1
60	(3.0)	(2.0)	(1.1)	(1.4)	(0.8)	(0.3)	0.2	0.4	0.4
70	(4.4)	(3.0)	(1.8)	(2.8)	(1.8)	(1.0)	(1.1)	(0.6)	(0.2)
80	(5.8)	(4.1)	(2.4)	(4.2)	(2.9)	(1.6)	(2.6)	(1.7)	(0.9)

[a]Yearly production from each plant is assumed to be 0.1 Quad. This corresponds to a production of 47,200 barrels per day for 365 days or a nameplate capacity of 52,450 barrels per day, assuming a 90 percent on-stream factor. Smaller or larger production amounts would scale in direct proportion. Plants are assumed to operate for thirty years.

[b]Price jump modeled as 100 percent increase followed by linear decline to return to long-run trajectory fifteen years after the price step.

jump event, the synfuel cost level, the year of the price jump event or initiation of the surge deployment effort, the lag time before production is obtained from surge deployment plants, and the discount rate.

Table 2-6 displays estimates of the net present value of each representative (0.1 Quad per year or about 50,000 barrels per day) synthetic fuel plant built as part of a surge deployment response to oil price jump events. Estimates are presented for stylized price jump events occurring in 1987, 1993, or 2000, as well as for a wide range of synfuel cost outcomes. Plants are assumed to begin production seven years after the price jump first occurs. Table entries show net present discounted value in 1980 of the oil/synfuel cost differential expressed in billions of constant 1980 dollars (1980$). Calculations assume a plant life of thirty years and a real social discount rate of 5 percent.

Net benefits or savings relative oil are positive entries. Where the discounted value of synfuel costs exceeds oil costs, net costs occur. These negative entries, or net production subsidies, are indicated by parentheses. Thus for some combinations of long-run oil price trajectories and effective synfuel costs (which reflect the oil import premium), surge deployment in response to a price jump event yields net subsidy costs rather than benefits. This occurs because synthetic fuels take a long time to deploy in response to an event and also because the economic life of the facility is long—thirty years. As a result, the underlying long-run price trend figures importantly in the net expected value, even for plants deployed in response to a major price jump event.

These tabulations of the benefits or costs from surge deployment plants reflect parametric variations in synfuel production costs and oil market prices. The third parameter—the subjective oil import premium—may be handled by treating it as a credit against synfuel production costs. Thus, for example, a $60 per barrel synfuel production cost would become a $50 per barrel "effective" synfuel cost level with the subtraction of $10 per barrel credit for the long-run oil import premium.

The values in Table 2-6 indicate potentially significant benefits for each additional surge deployment plant made possible by initial production experience. However, the net present value of each additional plant is clearly a strong function of the effective synthetic fuel cost level, underlying long-run oil price trajectory, and timing of the price jump event and corresponding surge deployment effort. The underlying long-run oil price trajectory is a particularly important determinant of the expected value of synfuel production. As the negative (or subsidy) figures in the table suggest, high synfuel cost expectations or low long-run oil price trends can result in negative expected values or net losses from plants deployed in response to even large deviations above longer-run underlying trends. The peak prices possible with some price jump events could lead to periods where oil costs substantially exceeded synfuel costs, but deployment in response to those transient high prices would not necessarily yield discounted life cycle savings. Because of the long time required to build synfuel plants and the long economic life of the facilities, initial savings relative to oil could be offset by later losses.

If surge deployment yields benefits, entries in Table 2-6 show that the benefits of each additional plant could plausibly equal about 0.5 to several billion dollars. Total benefits may be estimated by multiplying the appropriate estimate from Table 2-6 by the number of additional surge deployment plants possible. Total enhanced deployability benefits could potentially reach tens of billions of dollars (net present value).

Security Events. A surge deployment effort could be motivated by a sudden increase in the long-run security premium for oil imports even if the market price for oil continued to follow some smooth trend. Table 2-6 estimates would not be directly applicable to such a security scenario for surge deployment. For such surge deployment efforts, calculations utilizing the smooth long-run trends are appropriate.

Table 2-7 displays estimates of per plant benefits for security-motivated surge deployment efforts.[20] As for the price jump estimates, table entries present the net present value of benefits or costs for each nominal size surge deployment plant. Security events leading to initiation of a surge deployment effort are assumed to occur in various years as indicated on the table. Calculations assume plants begin production seven years after the surge deployment effort is initiated.

Table 2-7. Net Present Value for Each Surge Deployment Plant.[a] DOE-PFG Price Trajectories for Security Events in Year Indicated[b] (NPV of Savings or Subsidy in Billions of 1980$ at 5% Discount Rate)

Effective[c] Synfuel Cost Level (1980$/bbl)	Low Oil Prices			Medium Oil Prices			High Oil Prices		
	1987	1993	2000	1987	1993	2000	1987	1993	2000
20	2.3	1.9	1.3	3.7	2.9	1.9	5.1	4.0	2.6
30	0.9	0.8	0.6	2.3	1.9	1.3	3.7	2.9	1.9
40	(0.6)	(0.3)	(0.1)	0.9	0.8	0.6	2.7	1.9	1.2
50	(2.0)	(1.3)	(0.8)	(0.6)	(0.3)	(0.1)	0.8	0.8	0.6
60	(3.4)	(2.4)	(1.4)	(2.0)	(1.3)	(0.8)	(0.6)	(0.3)	(0.1)
70	(4.9)	(3.5)	(2.1)	(3.4)	(2.4)	(1.4)	(2.0)	(1.3)	(0.8)
80	(6.3)	(4.5)	(2.8)	(4.9)	(3.5)	(2.1)	(3.4)	(2.4)	(1.4)

[a]Yearly production from each plant is assumed to be 0.1 Quad. This corresponds to a production of 47,200 barrels per day for 365 days or a nameplate capacity of 52,450 barrels per day, assuming a 90 percent on-stream factor. Smaller or larger production amounts would scale in direct proportion.

[b]Figures are for plants starting production seven years after increased security premium motivating surge deployment effort. A 1987 event implies a 1994 plant start-up, and so forth.

[c]"Effective synfuel cost level" refers to net cost after subtracting the appropriate dollar per barrel value of oil import premium from total social cost of synthetic fuel production.

The parametric tabulations reflect a range of outcomes for the effective synfuel cost level. Benefits can be estimated for different values for synfuel production costs and oil import premiums by subtracting the value of the oil import premium (after its increase to reflect the security event) from the total costs of synthetic fuel production to yield an effective synfuel cost level.

Table 2-7 entries are similar in magnitude to those in Table 2-6. As with benefits for price jump surge deployment events, total benefits equal expected per plant benefits multiplied by the number of additional plants possible.

Conceivably, a surge deployment effort could be motivated by a price jump and security development occurring at the same time. In such a case, the per plant benefits could be estimated from Table 2-6 by simply adjusting the effective synthetic fuel cost level to reflect the increased value of the oil import premium.

The total value of potential enhanced deployability benefits is clearly sensitive to particular parameter assumptions. However, representative values from Tables 2-6 or 2-7 suggest that per plant benefits on the order of about one billion dollars (net present value) are plausible. If an additional ten plants are possible because of initial production experience, total benefits could reach tens of billions of dollars (net present value). As with the learning benefits for surge deployment, the insurance payoffs of enhanced deployment capability could be substantial.

Different assumptions about the key parameter values and uncertainties would yield different estimates for the magnitude of possible deployability benefits. The parametric display presented in the above tables is intended to allow individuals to assess benefits based on their own expectations for the value of key uncertain parameters.

Benefits from Avoiding Factor Cost Inflation. So far, this discussion has characterized enhanced deployability benefits from increasing the number of additional simultaneous plants possible in a surge deployment effort. However, one could argue that any production increment is possible at some cost. An economic description of expanding deployment capability would avoid absolute limits on the number of simultaneous plants possible. Rather, enhanced deployment capability would reduce the bottlenecks, supply shortages, and factor cost inflation that would otherwise be incurred with a deployment effort of a given size. Thus, enhanced deployability may result in "lower cost" plants in addition to the "more plants" benefits estimated above.

These cost reductions for surge deployment plants are separate and additive to the benefits of technology improvement or selection learning. With enhanced deployment capability, the entire group of surge deployment plants should encounter lower factor cost inflation thus reducing the per barrel production costs. The savings accrue over the lifecycle of plant operation and may be estimated in a manner entirely analogous to the estimates for learning benefits.

Total benefits from avoided factor cost inflation would depend on the size of the surge deployment effort and the extent to which costs for the entire group of plants built (not just the additional plants) were reduced. Both sources of benefits from enhanced deployment capability are important.[21] Reduced factor cost inflation will lower the costs for all plants built. The additional production volume possible may also be valuable.

Summarizing Total Benefits from Enhanced Deployment Capability. The experience and infrastructure development provided by initial synfuels production efforts should expand future surge deployment capability. Should an all-out deployment effort become necessary, more plants could be built, and factor cost inflation could be reduced from levels that would otherwise be expected. The size of the benefits of these two sources depends on assumptions for several uncertain parameters and scenarios.[22] Each additional plant possible in a surge deployment effort could yield benefits of about one billion dollars (net present value). Total benefits would depend on the number of additional plants made possible. The size and effectiveness of initial production efforts for increasing deployment capability should determine the number of additional plants possible and the benefits from "more plants." Benefits from avoided factor cost inflation may also be substantial. If a maximum surge deployment effort becomes necessary, total benefits from enhanced deployability on the order of tens of billions of dollars (net present value) are readily plausible.

The benefits of enhanced deployability, however, are realized primarily if maximum deployment capability must be exercised. Deployability benefits are important as insurance payoffs for surge deployment scenarios, not evolutionary growth of an industry. Moreover, even if a surge deployment event occurs, maximum deployment capability may not necessarily be exercised. It may be possible to completely displace oil imports with a volume of synfuels less than maximum deployment capability could provide. If maximum deployment capability were not utilized, benefits from expanded capability would be small. If basic deployment capability is large to begin with, the benefits of expanding deployment capability are less important. If the linkage between production experience and future capacity is strong, a small or moderate program may "buy" sufficient development of industry infrastructure. If the linkage is weak, however, deployability increases may be small, and larger initial production efforts may be needed to substantially expand deployment capability.

The benefits of enhanced deployability, while potentially totaling tens of billions, are not certain to be obtained. Surge deployment may not be necessary nor maximum deployment capabilities utilized, even if an event occurs. Development of the industry infrastructure is a potentially important insurance payoff of initial synfuels production experience, however, and enhanced deployment capability provides a rationale for special attention to synfuels. Infrastructure effects are relatively specific to energy options such as synthetic fuels with their long lead-times for the development of specialized skills, experience, and capabilities. Moreover, the benefits of enhanced deployment capability accrue to other firms and the nation as a whole, rather than entirely to the private firms whose pioneering projects build the infrastructure which expands options for others.

Improving the Growth Basis for Surge Deployment

A surge deployment effort would concentrate in time a large increase in the level of economic activity associated with the design, construction, and operation of synfuel facilities. If, in any given locale, this induced economic activity reflects primarily a shift away from other activities to synthetic fuels, then the socioeconomic impacts of synfuels deployment may be minor for that locale. However, if synfuel production efforts in a given locale represent a significant absolute increase in the aggregate level of economic activity, substantial socioeconomic adjustments may be necessary. The magnitude of the socioeconomic costs or externalities associated with these adjustments are related to the magnitudes of change and rates of growth.

Many of the difficulties of rapid growth stem from the relative rate of growth rather than the absolute amount of growth.[23] Consequently, areas with relatively high levels of existing economic activity and population can make the adjustments required to support synthetic fuel deployment more easily than sparsely developed locales.

Initial synfuel production efforts could yield social benefits by improving the "growth basis" for a surge deployment effort, thus reducing the socioeconomic costs of surge deployment efforts. For synthetic fuels in particular, the potential value of improving the growth basis for surge deployment is magnified because the extraction costs and distribution of U.S. coal resources make it economically advantageous to locate some synthetic fuel production facilities near the western coalfields. Extraction costs for western surface mined coal are substantially lower than costs for surface or underground mined coal from the eastern coal reserves. The impacts of aggressive synthetic fuel deployment could be particularly significant in relatively sparsely populated western regions.

Initial synfuel production efforts could increase economic activity in key production regions. Moreover, unlike crash deployment efforts, they could be structured, paced, and executed with attention to measures that could mitigate potentially adverse socioeconomic or environmental impacts. Well-planned initial production activities would improve the socioeconomic infrastructure for rapid growth and reduce socioeconomic costs in case of future surge deployment.

An improved growth basis could translate ultimately into more and better plants possible in a surge deployment effort. Well-executed initial production efforts could improve deployment prospects by establishing both an improved community infrastructure and political climate to support major expansion. For example, community attitudes and physical facilities to support construction of a second synthetic fuel facility may be favorable if an initial project has gone well. Lower resource costs for synfuel production may also be obtained as the high local price inflation, economic dislocations, poor labor productivity, and inefficiencies associated with boomtown effects are reduced.

Quantitative assessment of the benefits from an improved growth basis is difficult. Many of the values involved are intangible matters of subjective judgment—such as reduced boomtown character and crime, or enhanced political climate. This study does not develop estimates for these potential growth basis benefits. Nevertheless, given the strains that a surge deployment effort could place on the socioeconomic and political fabric of particular regions and the nation as a whole, pioneering synthetic fuel production efforts should consider how initial projects can be structured and executed in ways that improve the basis for later production expansion in key producing regions. Should emergency deployment become necessary, an improved growth basis should allow larger and faster production increments, lower costs, and less adverse socioeconomic impacts, as well as easier tradeoffs among important social and political values (Seidman 1980).

INFORMATION AND CAPABILITIES AS STRATEGIC INSURANCE

This chapter has outlined a taxonomy for the benefits of initial synthetic fuels production experience. The taxonomy distinguishes among three broad categories of potential benefits.

The assessment of "production benefits" identifies some social values that could be associated with any option for reducing oil imports. The value of these benefits as represented by the oil import premium, however, may not be large enough to offset the potential production costs for synfuels as compared to the market price of imported oil. Moreover, neither the oil import premium nor energy security concerns provide strong reasons for giving special attention to synthetic fuels over any number of alternative ways to reduce oil imports. Thus, potential production benefits provide little rationale for special emphasis on synfuels or guidance for the design of cost-effective initial production efforts.

In contrast, two other categories of benefits—"information benefits" and "surge deployment" benefits—relate more specifically to synthetic fuels and provide more guidance for assessment and design of initial production efforts. The substantial uncertainties surrounding synthetic fuels production costs and other issues magnify the potential information benefits from initial production experience. Reduced uncertainty could improve major policy and long-term investment decisions resulting in potential cost savings totaling tens of billions of dollars (net present value).

Information on costs and production prospects for major oil substitutes such as synthetic fuels could also affect oil pricing. The direction and magnitude of potential oil pricing effects are speculative. They occur only if the unknowable existing expectations and behavior of the actors who influence oil prices are changed as a result of information from initial synfuel production experience. Potential losses or gains could be large, on the order of hundreds of billions of dollars. A key attribute of these general information benefits is that most of the uncertainty can be reduced via only a few initial projects.

The third major category of benefits—surge deployment benefits—provides additional rationale and useful guidance for design of initial synthetic fuel production efforts. Given the potential for rapid change in the market price or security liabilities of imported oil, production experience can yield the tangible social benefits of lower cost plants (learning) and more plants (enhanced deployability) in a surge deployment effort. These benefits are primarily important if some event motivates an emergency deployment effort. They are the insurance payoffs of the improved surge deployment capabilities obtained via the insurance premium of the production subsidies that may be required to promote initial production projects.

The total insurance payoff from various sources of surge deployment benefits can be substantial. For a surge deployment effort comprising about thirty plants (1.5 *MMBD*) in the 1990s, selection learning combined with technology improvement learning could readily yield benefits totaling tens of billions of dollars net present value (*NPV*). Enhanced deployability benefits could approach billions of dollars (*NPV*) for each additional surge deployment plant possible. Thus, total payoffs could reach several tens of billions of dollars.

However, these insurance payoffs are obtained primarily in the unhappy event that a surge deployment effort becomes necessary. The "expected" (or probability-weighted) value of surge deployment benefits should be adjusted to reflect the odds that a surge deployment event will occur. Benefit-cost assessments for initial synthetic fuel production efforts should weigh the expected costs for initial production subsidies (the insurance premium) against the probability-weighted value of surge deployment benefits (the insurance payoffs).

Because surge deployment benefits provide rationale for special attention to synthetic fuels, they are a central concept for this study. Many of the improved capabilities stemming from initial production experience yield benefits to society without adding revenue to the balance sheets of the pioneering projects. For example, many technology improvements cannot be protected by secrecy or patents. The training of synfuel design firms or construction labor yields future savings to the nation as a whole without commensurate benefits to the private firms undertaking a pioneering project. Moreover, the benefits are obtained via large investments that pay off in unfortunate and, hopefully, low probability events. Accordingly, market opportunities and incentives alone may not lead to levels of initial production experience (or insurance protection) appropriate for the nation as a whole. Special government efforts may be warranted.

The parametric estimates presented here provide some quantification of the possible magnitude of various sources of benefits from an initial synthetic fuels production program. Chapter 3 follows with a corresponding taxonomy and parametric assessment for important categories of program costs. Together these chapters, with both their qualitative frameworks and quantitative estimates, help illuminate "Whether and why?" special investments with synfuels may be appropriate. Answers to these questions depend on individual subjective expectations for key uncertain outcomes, such as synfuel costs and the likelihood of surge deployment insurance events. The framework for benefits and costs also provides the basis for considering the "How large?" question, addressed in Part II of this study. Finally the "strategic capabilities as insurance" benefits of a initial synfuels production effort define important program purposes, suggesting issues and approaches for doing better with these major national investments.

NOTES

1. The literature of serious assessments of the oil import premium is developing rapidly. One of the best and most systematic assessments is presented by William Hogan in "Import Management and Oil Emergencies" (Deese and Nye 1980). Hogan shows that plausible variations in world view as well as energy market and policy responses could generate a range for the oil import premium from $2 to $40 per barrel. For other significant efforts toward sound assessments see Department of Energy (1980c), Nordhaus (1980), and Rowen and Weyant (1980).
2. For example, a constant subsidy of $10 per barrel for one plant producing 0.1 Quad per year beginning operation in 1990 yields discounted life cycle

subsidy costs of $1.8 billion at a 5 percent discount rate and $0.6 billion at a 10 percent discount rate.
3. The term "effective shortfall" is used to reflect the possibility that storage or other supplies could be used to replace part of disrupted supplies, thus reducing the effective size of any disruption of normal supplies.
4. The calculation is straightforward: $1/bbl × 365 days/year × 1,000,000 bbl/day = $365 million per year, which totals with discounting over 20 years, to 12.4 × $365 million, or $4.5 billion. 12.4 is the accumulating discount factor for a constant stream of values over 20 years at a 5 percent discount rate.
5. Synthetic fuels are commonly viewed as a backstop to rising fuel prices. However, possible constraints on maximum feasible production rates, as well as increasing marginal costs with increasing production levels, probably keep synthetic fuels from functioning as a true backstop.
6. The model used is described in ICF, Inc. (1979a) and reflects the approach of Salant (1976). The experiments are summarized in Powell (1979) as draft analysis developed in the Department of Energy's Office of Policy and Evaluation during 1979. As an illustrative calculation of the potential magnitude of oil pricing benefits or costs, note that a $1 per barrel shift in oil prices over 8 *MMBD* of imports aggregates with discounting to a net present value (using a 5 percent discount rate) of about $58 billion.
7. For example, if some information outcome changes the expectations of oil pricers and causes a $100 billion *NPV* change in the oil price trajectory, a 10 percent probability for changed expectation combined with a 10 percent belief that a particular economic model predicted oil pricing behavior would still produce an expected cost of benefit on the order of $1 billion.
8. The economic term "externalities" refers to social costs or benefits that are not reflected or readily capturable as part of marketplace transactions. For example, increased crime rates in small towns near synfuel projects represent a cost to society that is not readily reflected in the costs for constructing and operating a plant. For more information, see most basic economics texts, including Musgrave and Musgrave (1973) or Mansfield (1970).
9. This is not to suggest that government is better in general at assessing probabilities and futures. Indeed, the profits (or losses) realized from correct (or incorrect) assessment of the future give private sector individuals strong incentives to do well. Rather, the suggestion is that, for the partially politicized energy market in particular, intelligence information and the role of foreign and domestic policies may put government in a slightly better position to assess and even influence the probabilities and descriptions for potential surge deployment scenarios.
10. Much of the experience has been acquired only recently via initial work on projects within the United States. Several U.S. companies have indirect experience via the participation in the SASOL project in South Africa.

11. For a good review of the learning curve concept and the context for its application, see Bodde (1976).
12. David L. Bodde first suggested this representation of surge deployment events and learning curves.
13. The debate about learning rates ranges widely from very small percentages to over 20 percent. As discussed in Merrow (1978), learning with capital-intensive technologies may be smaller than learning expected with relatively labor intensive technologies. Moreover, many of the process operations and equipment in a synthetic fuel plant may already be commonly applied in other industries. Thus, the learning potential for major elements of a plant may be quite limited. Accordingly, high learning rates for new process elements should not be applied to the whole plant or total production costs including feedstock costs. Well-characterized low estimates for the learning factor are represented by Merrow (1978:54) at about 2 percent. The optimistic end of the range may be represented by Hirschmann (1964), who suggests that a 20 percent learning curve may be applicable to capital intensive chemical processing technologies also. A few empirical studies are available. One recent study of coal-fired power plant construction reported a learning curve factor of about 8 percent. See Ostwald and Reisdorf (1979). Given the caveats about applying learning curves primarily to capital costs for innovative process elements, a range of 2 to 15 percent covers many expectations.
14. For purposes of this study, a "nominal," "representative," or "full size" synfuel plant produces 0.1 Quad per year. This corresponds to about 47,200 barrels per day on average, or a production capacity of about 52,000 barrels per day, assuming a 90 percent on-stream factor. Where quantitative precision is unnecessary, a nominal plant may be viewed as the commonly considered 50,000 barrel per day facility. Twenty nominal facilities are needed to yield one million barrels per day.
15. Some interrelationships, however, should be considered in interpreting the information from initial synthetic fuels production experience. For example, technology cost comparisons prior to accounting for technology improvements could be different after learning effects are considered. Also, synthetic fuels processes share a sufficient number of common elements that advances with one technology may be largely applicable to other technologies as well. Technology improvement learning makes it more difficult to realize selection learning.
16. This observation is made forcefully by Cameron Engineers (1979) as well as by several Rand Corporation studies. See the discussion of factor cost inflation in Chapter 3.
17. Building the industry infrastructure is analogous to an "external economy of production." These effects are recognized by economists to yield positive

benefits to society that cannot be wholly captured by the private firm undertaking the production project. See, for example, Baumol (1972:392–393).

18. The physical analogy is to the critical mss required to initiate a nuclear chain reaction. An industrial analogy is the minimum market potential required to induce capacity expansions.
19. The rate of supply and demand responses varies with technology, economic growth, the life of equipment, the decisions of investors, and so on. Many adjustments can take place in one or two decades. For perspective on demand adjustments and their rate, see Hogan (1979).
20. The term "security-motivated" is not meant to suggest any special energy security role for synthetic fuels. Rather, it refers to an increase in the long-run oil import premium due to a sudden increase in the security component of the premium. Such an increased premium should be consistently applied to all energy production and consumption alternatives.
21. The decision analysis assessment in Part II reflects both sources of benefits from enhanced deployment capability. An absolute barrier limits production increments in any given period to some maximum, depending on base deployment capability, previous production levels, and the linkage factor (*DIF*) between past production experience and future deployment capability. However, the expected level of factor cost inflation is also reduced for any given production increment less than the maximum possible.
22. In addition to these surge deployment benefits, improved national capability to respond quickly and in large scale with major alternatives to imported oil could possibly yield some general benefits associated with oil pricing behavior. Long-run price trajectories might be moderated very slightly, and large short-run price increases might be deterred. Although any oil pricing benefits of enhanced deployment capability are very speculative, such benefits, if they exist, may be realized without actual occurence of a surge deployment effort. See Appendix II-C of Harlan (1981) and Hoffman and Kennedy (1981) for more discussion.
23. For a good bibliography of the literature on boomtowns, see Office of Technology Assessment (1980:470).

3 COSTS FOR INITIAL SYNFUEL PRODUCTION EFFORTS
Basic Elements and Assessments

INTRODUCTION AND OVERVIEW

Chapter 2 characterized potentially substantial benefits from initial synthetic fuels production efforts. Such benefits provide positive answers as to "Why?" a special synfuels program might be warranted. The costs of initial production projects, however, may counterbalance the benefits. Comparisons of benefits and costs are needed to gain real insight on the question of "Whether?" an initial synfuels program is worthwhile. This chapter presents basic elements and assessments for a taxonomy of costs for initial synfuel production efforts.

A single commercial scale synthetic fuel plant requires multibillion dollar capital investments. There are opportunity costs for applying these resources to synthetic fuel production rather than to other economic activities. Initial production projects will incur net costs unless the value of synfuel products exceed the costs of production. Two key factors in this comparison—synfuel production costs and oil price trajectories—are uncertain. As a result, the size of net costs or savings for each barrel of synfuel production remains uncertain.

The uncertainty regarding the straightforward, balance sheet profitability for initial synfuels production projects complicates terminology. As a convenient convention, this taxonomy assumes that the production costs will exceed product revenues for initial commercial coal-based synfuel projects. Over the lifecycle of initial plants, balance sheets will show net losses relative to oil. The possibility of net savings relative to oil is recognized, but we assume that initial production projects will require net production subsidies. Without this assumption, it is more likely that private firms will invest in synfuels production in order to realize market opportunities for profit, thus making special government incentives or programs unnecessary.

This study examines three major categories of costs for an initial synfuel production program. The first category is the net production subsidies for a single representative project. These costs consist of the production costs for synthetic fuels compared to a price trajectory for the crude oil or petroleum products displaced by the synfuel products. Synfuel production costs are examined to show why they are uncertain and could vary over a wide range. Future oil price trajectories are even more uncertain and difficult to predict. Accordingly, net costs for production subsidies are estimated parametrically for a range of assumed synfuel costs and oil price trajectories.

A second major category of costs is related to the aggregate size of a group of simultaneous projects. Costs for a program comprising many individual plants could exceed the simple multiplication of the number of plants times the expected cost of a single representative plant built by itself. High levels of demand for certain key professional skills, specialized labor skills, and material components could induce substantial price increases for some key inputs or "factors of production" required to design, manage, construct, and operate synfuel projects. Even where prices did not increase, bottlenecks, scheduling difficulties, and other inefficiencies could increase effective plant construction costs. This taxonomy applies the term "factor cost inflation" to this important category of costs.

The issues and phenomena underlying factor cost inflation are complex. Quantitative assessment is difficult but important. Illustrative estimates of the possible magnitude of factor cost inflation are developed by analogy to the 1973-74 period of "hyperinflation" in chemical plant construction costs. The experience of that period provides practical evidence of the substantial real cost increases that may be incurred when the level of construction effort strains available supplies and capabilities (Savay 1975).

The environmental and social impacts of synfuel production are the third important category of costs. The costs for designing and executing a synfuel project that complies with all applicable environmental requirements should be included in a complete specification of synfuel production costs. Even when all environmental standards are met, however, some additional environmental or socioeconomic costs may be incurred. For example, the impact of a synfuels project on a scenic view is an environmental cost for which no standards exist. Also, the rapid growth in economic activity and population associated with synfuels projects may strain the social fabric in some locales. Socioeconomic adjustments and intangible costs may be inevitable.

These environmental and social costs add to the tangible production costs to increase the effective cost level for synthetic fuels. Such increases in the effective costs counterbalance the effective reductions in synfuel costs obtained when synfuels are credited with a value for the oil import premium. Assessment of both environmental and socioeconomic costs as well as the oil import premium involves many subjective values. This study does not attempt an elaborate description or quantification of potential environmental or social costs. While these costs should clearly be recognized and addressed in policy formulation and program design, they are as uncertain and subjective as some elements of the oil import premium which they counterbalance. Fortunately, numerous other studies are giving careful attention to these issues. Integration of those assessments into program design should be an important complement to the perspectives offered by this study.

Review of these three major categories of potential costs—net production subsidies, factor cost inflation, and environmental/social costs—suggests two major observations for the costs of initial synthetic fuel production efforts. First, potential costs are inherently uncertain and could vary over a large range.

Second, costs are very likely to increase as the size of an initial production program increases. In particular, the magnitude and likelihood for factor cost inflation is strongly related to the aggregate size of an initial production program.

This chapter reviews the major components of cost supporting these observations. Net production subsidies are described by characterizing synfuel production costs, oil price trajectories, and the appropriate comparison of those quantities over the lifecycle of a representative synfuels project. Following estimates of production subsidies for a single representative plant, the phenomena underlying factor cost inflation for a program of many projects are reviewed. A concluding section summarizes implications of this taxonomy and assessment for the evaluation of "Whether?" initial synfuels production programs are likely to be worthwhile.

NET PRODUCTION SUBSIDY COSTS

The net costs for synfuel production subsidies depend on two major uncertain factors—synfuel costs and petroleum prices. Synthetic fuel production costs should be carefully defined to include the full costs for the resources applied to the development, design, construction, and operation of a synfuel plant. Production costs should be broadly defined to include costs of complying with all applicable environmental requirements. Costs for new community and economic infrastructure devoted to synfuel production should be considered. Such a "real resource cost" accounting should reflect the full opportunity costs of applying the nation's resources to synthetic fuels production rather than some other purpose.

The relevant oil prices against which to compare synfuel production costs are also difficult to define. Production subsidies should be estimated over the operating lifecycle of a synfuel plant. Long-run trajectories for the market price of crude oil are fundamentally uncertain. Moreover, the market value of synfuel products depends on the market value of the petroleum products displaced.

Estimates of net production subsidies are based on simple calculations with these uncertain factors. Rather than attempting to develop particular estimates, this study characterizes plausible ranges of variation and develops parametric estimates of net production subsidies.

First, a range of variation for synfuel production costs is described. Then, alternative example long-run crude oil price trajectories are reviewed briefly along with background on the value relative to crude oil ("form value") of various petroleum products. These descriptions provide the basis for parametric tabulations of a range of possible net production subsidy costs for a representative initial synthetic fuel project. Parametric estimates for alternative oil price trajectories, initial production dates, and discount rates are presented

Ranges for Synthetic Fuel Production Costs

Synfuel production costs are a key parameter in any cost-benefit assessment. It is important to understand why basic synfuel production costs are both understandably uncertain and potentially high. Detailed review of technology flow sheets is not necessary to appreciate the key elements of synfuel production costs. The technical viability of a number of alternative processes is well known. The key uncertain issue remains the cost of production.

Synfuel production costs have received considerable attention over the past decade. Although inflation complicates the story, cost estimates for synfuels seem to have risen as rapidly in real terms as oil prices. As we will see, one reason for the apparent cost increases is that the engineering studies and cost estimates are becoming more detailed as actual investment decisions come closer to realization.

General Components of Synfuel Production Costs. Understanding of synfuel production cost estimates is aided by separating total costs into several major classes: plant capital costs, operation and maintenance costs, feedstock costs, capital charge rate, plant capacity factor, and associated infrastructure costs. "Plant capital costs" are generally the largest component of costs. They include the direct investment outlays for hardware, construction labor and services, engineering services, land, and so forth, required to build the production facility. Depending on accounting definitions, plant capital costs may include the interest charges on capital expended for plant equipment before the plant begins operation. "Operation and maintenance" (O&M) costs include the costs required to operate and maintain a facility once construction has been completed and the plant is producing in a normal commercial operating mode. "Feedstock costs" reflect the costs for coal input to the plant for transformation into synfuel products.

These three basic cost elements translate into per unit synfuel production costs via two additional factors. The "capital charge rate" reflects the annual revenue per dollar of investment required to repay capital costs and yield an acceptable rate of return for money invested now which provides future revenue.

The "plant capacity factor" indicates the quantity of products produced from a synfuel plant over a given accounting period compared to the nominal amount of production possible from that plant if it were producing at full capacity for the entire period. Plant throughput determines the total volume of products over which fixed costs for capital investment and operation and maintenance are spread. Since per unit costs equal total costs divided by the total amount of production, plant performance has a strong effect on the per barrel production costs.

A full cost accounting may properly include some additional costs that are a necessary adjunct to a synfuels project but that are not directly at the plant site or linked to the production technology. These "associated infrastructure costs" include towns and community facilities for project personnel as well as the transportation facilities required for a project, such as rail or pipeline facilities to deliver supplies or move products to markets. The sections below characterize representative values and approximate ranges of variation for each of these major components of synfuel production cost.

Plant Capital Costs. Plant capital costs have been the focus of many recent assessments of synthetic fuel production costs because they are the largest component of total costs. As a planning matter, it seems possible to estimate plant capital costs without actual construction experience. Many studies have estimated plant capital costs on paper using established engineering-economic techniques. Engineering studies can define the amounts of equipment, hardware, materials, construction labor, engineering services, and other resources required to build a synfuel plant (Braun 1976; Cameron Engineers 1979; Oak Ridge National Laboratory 1981). If a construction schedule is known, total plant capital costs, including interest costs during construction, may be calculated.

Several phenomena, however, combine to make total plant capital investment costs relatively uncertain. First, the costs of building a plant, even using a well-known process, depend to some extent on the particular site for a plant. Different coals may require different equipment sizes or environmental control techniques. Different climates may change equipment efficiencies or require design modifications. The land cost and terrain itself affects plant capital costs. Estimates of plant capital costs for general, unspecified sites are inherently uncertain, and, given a tendency to assume convenient site characteristics, they frequently underestimate costs.

Second, the actual amounts, arrangement, and organization of the equipment, labor, and other resources required to build a complex multi-billion dollar coal processing facility require detailed and expensive planning. Especially for first of a kind projects, the complexity of a production plant using an innovative process is often not fully appreciated until a detailed engineering design is complete. Total costs are often uncertain until construction is complete and initial operations have clarified whether major equipment modifications are needed. Detailed engineering design requires definition of an actual project site and can cost tens of millions of dollars. Because few synthetic fuel projects have progressed to this stage, estimates of capital costs should reflect broad ranges.

The Rand Corporation has conducted empirical studies of escalation in cost estimates for energy process plants (Merrow, Phillips, and Myers 1981; Merrow, Chapel, and Worthing 1979). Those studies demonstrate a strong tendency for estimates of plant cost to understate plant complexity and capital cost until a definitive, detailed engineering design for budgeting the plant construction is available. Depending on the sophistication and effort behind preliminary estimates, actual plant construction costs may exceed earlier estimates by a factor of as much as two or three. Many commonly cited estimates of synthetic fuel plant capital costs are preliminary estimates for generic process designs not associated with a particular site. Such estimates have a wide uncertainty range (about 25 percent) and usually underestimate plant complexity and capital cost by substantial percentages.[1]

This phenomena—increasing realization of plant complexity with increasing design effort as a plant comes closer to actual design and construction—is one reason that estimates of synfuel product costs have been steadily increasing. As oil prices have increased, public and private sector entities have been increasingly willing to make the multi-million dollar investments in detailed planning and design needed for a good estimate of costs as a basis for multi-billion dollar investment decisions. As the design effort becomes more detailed more plant complexities are recognized, and cost estimates increase

Third, even if one accurately knew the amounts of inputs required to build a synthetic fuel plant (such as major equipment items, concrete, piping, and other physical materials, construction labor with general or specialized skills, and professional engineering and other services), the costs for obtaining these goods and services depend on specific circumstances. For example, the price for certain major equipment items may depend on whether suppliers charge list price or discount items according to usual practice. The amount of discount may depend on market demand, including demand from other sources such as construction of chemical plants in this country or abroad. Costs for other inputs could be influenced by developments that depend on government decisions. For example, the price and availability of construction labor or reinforcing steel bar ("rebar") for construction of synthetic fuel plants in the west may depend on the progress of the MX missile system. Construction labor productivity depends on specific local situations. Thus, even if amounts of inputs are relatively well-known, potential price or productivity variations cause lingering uncertainty for total plant capital costs.

Finally, variations in project construction schedules may result in uncertainty for the "interest during construction" component of plant capital costs. Delays from a planned schedule can significantly increase costs for interest during construction. Such delays could arise if construction difficulties are encountered, if key equipment items do not arrive at the plant site on schedule, or if regulatory, strikes, or other procedural problems are encountered.

Taken together, these phenomena explain some of the wide variation and recent increases in synfuel cost estimates. Varying rates of inflation for general prices and construction costs further complicate comparisons of estimates made at different times and with different assumptions or engineering standards (Cameron Engineers 1979:173). This analysis need not repeat or assess the many estimates available. The major point—that significant uncertainty exists for good reason—can be made without detailed examination of particular estimates.

A few responsible estimates may be used to benchmark an approximate cost range. Table 3-1 displays three serious estimates of capital costs for coal-based synthetic fuels. Since plant sizes vary, total capital investment costs are expressed as dollars of investment per daily barrel oil equivalent production capacity. No attempt has been made to see if these estimates were consistent in their treatment of various components of capital cost. The estimates are for plants that produce liquids or gases from coal via an intermediate coal gasification step. Among the estimates in Table 3-1, the Fluor estimate probably most fully reflects full plant complexity, since it is based on the actual detailed design and construction experience from the Sasol II facility in South Africa. Other estimates are based on less detailed engineering design.

As a representative benchmark value for quantitative assessments using plant capital costs, this study will utilize an estimate (near the Flour estimate) of about $60,000 (1980$) plant investment cost per daily barrel oil equivalent production capacity. Sensitivity analysis within a range of about 30 percent around this reference value is appropriate.

Operation and Maintenance Costs. Operation and maintenance (O&M) costs are an important element of total production costs. O&M costs include expenditures for such items as operating labor and utilities (water, electricity, etc.), catalysts and other chemicals, materials and labor for plant maintenance, property taxes and insurance, and general plant administrative overhead. A portion of O&M costs (such as property taxes and insurance) are fixed yearly amounts related to the total plant investment rather than to the level of plant production. Other O&M costs (such as catalyst and chemicals) vary with plant production level. Total O&M costs are relatively uncertain. They can vary with total plant investment, the particular process design, and plant production level.

Little practical experience is available to provide a firm basis for estimating the O&M costs for large coal processing facilities. Nevertheless, many O&M cost items are commonly estimated as percentages of total plant facilities investment. Table 3-2 displays estimates for fixed O&M costs as a percentage of plant capital costs.[2] Fixed O&M costs exclude utilities, catalysts, and chemicals and do not vary greatly with plant production volume.

Table 3-2 shows that fixed O&M cost items are typically estimated to be about 7 to 10 percent of plant fixed investment. If one assumes that variable O&M costs for consumable utilities such as electricity, water, catalysts, and chemicals are approximately offset by revenues from by-product sales,[3] net O&M costs are a strong function of the capital costs per daily barrel of production capacity. A calculation that assumes yearly net O&M costs of about 6 to 10 percent of plant capital investment costs yields an O&M cost estimate of about in the range of $10 to $20 per barrel.[4] The partial

Table 3-1. Representative Estimates of Plant Capital Investment Costs for a Coal-Based Synfuel Plant.

Estimate Source	Plant Capital Costs	Comments
	(1980$ Per Daily Barrel Crude Oil Equivalent of Production Capacity)	
		Sasol-II modified for U.S. western site
Fluor estimate for Fischer-Tropsch gasoline[a]	$62,000	$3.6 billion mid-1979$ increased 10% to 1980$
		65,000 bbl/day production capacity
		Includes premiums for special labor camps and overtime
Liquefaction Technology Assessment (LTAS) by Oak Ridge National Lab for DOE[b]	$51,000 to $79,000[c]	Estimates in 1979$ updated 10% to 1980$
(Coal to methanol or gasoline)		High end of range for maximum gasoline production
Cameron Engineers (estimate for indirect liquefaction or high Btu gas from coal[d])	$38,000 to $47,000	1979$ estimates updated 10% to 1980$
		Lower end of range more representative of methane from coal

[a]Estimates derived from Fluor Corporation, 1979, *A Fluor Perspective on Synthetic Liquids: Their Potential and Problems*, pp. 28-29.

[b]Estimates presented in Oak Ridge National Laboratory, 1981, (prepared for U.S. Department of Energy) *Liquefaction Technology Assessment—Phase 1: Indirect Liquefaction of Coal to Methanol and Gasoline Using Available Technology*.

[c]Lower end of LTAS range is for coal to methanol or coal to gasoline with methane co-product.

[d]Cameron Engineers, 1979, "Overview of Synthetic Fuels Potential to 1990," in *Synthetic Fuels: Report by the Subcommittee on Synthetic Fuels of the Committee on the Budget—U.S. Senate*, p. 23.

Table 3-2. Factors for Estimating Fixed Operation and Maintenance Costs (estimates for coal to liquids or methane via intermediate gasification with Lurgi technology).

O&M Cost Item	Annual O&M Cost Items as Percentage of Plant Facilities Investment		
	Braun[a]	LTAS[b]	General SRI Factors[c]
Maintenance materials	1.2%	1.2%	1-3%
Maintenance labor	1.8%	0.3%	1-2%
Operating labor and supervision	0.7%	0.3%	NA
Administrative expenses, payroll burden, support labor	1.5%	2.6%	2%
Property taxes and insurance	2.7%	2.8%	2.5%
Totals	7.9%	7.2%	6.5% to 9.5%
Total plant facilities investment costs	Braun: $1060 Million (1976$) LTAS: $2110 Million (1979$)		

[a]Braun estimate from "Factored Estimates for Western Coal Commercial Concepts," ERDA/AGA Contract Report, October 1976, p. 73.

[b]LTAS estimates derived from Oak Ridge National Laboratory, 1981, (prepared for U.S. Department of Energy. *Liquefaction Technology Assessment—Phase 1: Indirect Liquefaction of Coal to Methanol and Gasoline Using Available Technology.*

[c]The General SRI Factors were not applied to an estimate for a particular technology. Rather, they were generic factors reported in a presentation at the First Annual Client Conference for the Synthetic Fuels Program of SRI, International in March 1979.

linkage of O&M costs to uncertain capital costs increases the uncertainty regarding O&M costs.

Feedstock Costs. Despite the abundance of coal in the United States, the net costs for coal feedstock for synfuel production is subject to considerable variation and uncertainty. Costs for coal extraction are sensitive to the characteristics of a particular coal deposit as well as to the mining techniques and environmental or safety requirements. Delivered coal costs depend on the location and transportation linkages for a synfuel project relative to its

Table 3-3. Net Feedstock Costs for Varying Coal Costs and Coal-to-Products Conversion Efficiency (1980$ per barrel of product assuming 5.8 million Btu per barrel of product).

Coal Cost	Conversion Efficiency (output energy in products divided by energy in coal input)						
	45%	50%	55%	60%	65%	70%	75%
(1980$/mmBtu)							
0.50	6.4	5.8	5.3	4.8	4.5	4.1	3.9
0.75	9.7	8.7	7.9	7.3	6.7	6.2	5.8
1.00	12.9	11.6	10.5	9.7	8.9	8.3	7.7
1.25	16.1	14.5	13.2	12.1	11.2	10.4	9.7
1.50	19.3	17.4	15.8	14.5	13.4	12.4	11.6
1.75	22.6	20.3	18.5	16.9	15.6	14.5	13.5
2.00	25.8	23.2	21.1	19.3	17.8	16.6	15.5

sources of feedstock supply. Net feedstock costs depend on delivered coal costs as well as on the coal-to-products conversion efficiency actually achieved. Coal-to-products efficiency for a given technology can vary with the technology operational mode and often does not meet design expectations.

Coal cost divided by coal-to-product conversion efficiency yields the feedstock cost per unit of product. A range of parametric estimates for coal feedstock costs may be developed by varying two basic parameters—the coal cost and the coal-to-product conversion efficiency. Table 3-3 displays parametric estimates for a range of coal costs and conversion efficiencies.

Table 3-3 shows that feedstock costs could vary widely from only a few dollars per barrel of product to over $20. Representative costs within this range may be estimated by reference to the existing cost for coal delivered to major coal-fired electric generating stations. Near the western coal fields these costs are about $0.75 per million Btu; near the Ohio Valley coal prices are about $1.25 to $1.50 per million Btu.[5] Conversion efficiency depends on the technology used and its actual performance. A representative value for processes likely to be applied commercially in the 1980s decade is about 55 percent.[6] Although highly case-specific, a reasonable range for net coal feedstock costs is about $7 to $15 dollars per barrel oil equivalent.

Capital Charge Rate. Capital costs for plant investment incurred at the beginning of a project yield revenue from the product sales over the lifetime of plant operation. The "capital charge rate" is a summary parameter translating a dollar of capital investment into an annual revenue stream that pays back the original capital outlay and provides a return on the capital investment. For example, a 10 percent capital charge rate means that, for every dollar of initial investment, revenue equal to $.10 must come every year in order to repay the original investment plus the time value of money. The capital charge rate summarizes the effective outcome for a wide range of financial factors, including interest rates, taxes, depreciation schedules, rate of return on debt and equity, and the relative amounts of debt and equity for a project.

With so many degrees of freedom summarized by one factor, it is not surprising that there is a wide range of possible values for capital charge rates. Capital costs are a major component of total synfuel production costs. Consequently, changes in financial assumptions and capital charge rates can cause projected per barrel product costs to vary widely, even if plant capital investment, O&M costs, and feedstock costs are the same. Views of the appropriate capital charge rate depend partially on subjective judgments, such as risk assessments. The spectrum of values ranges widely from 6 percent for long-lived, low-risk projects using utility financing and predominantly debt capital, to over 25 percent for relatively short-payback, high-risk projects financed largely with equital capital (Tebbetts 1978).

This study does not attempt to explore the complex issue of the appropriate financing structure and capital charge rate for assessing synfuel production costs. Rather, it assumes a standard capital charge rate recommended by some Department of Energy guidance on standardized cost estimates (Tebbetts 1978; Ray 1979). The capital charge rate assumed reflects a stream of annual revenues remaining constant in real terms over the 30-year economic life cycle of a synfuels project. DOE standardized planning guidance recommends a capital charge rate of 0.13 for thirty-year projects requiring five years to construct (Ray 1979:24). This value—13 percent per year—will be the standard capital charge rate applied for the estimates of real social resource costs in this study. When this capital charge rate is combined with the assumed nominal plant investment cost of $60,000 per daily barrel of production capacity, the resulting calculation yields a per barrel capital cost of about $24 1980 dollars per barrel.[7]

Plant Capacity Factor. Most estimates of per unit production costs for synthetic fuels assume that shortly after plant start-up, a synfuel production facility will produce 90 percent of its design capacity in a given year. Ninety percent is a common assumption for an expected capacity factor, since all or part of a plant must be shut down for regular maintenance.

The production rate actually achieved, however, is often significantly different than the standard assumption. Operational difficulties, design inadequacies, or other problems could cause average throughput or production volumes to be substantially less than those assumed for plant design. Alternatively, design conservatism, overdesign, or operational improvements could yield production rates that exceed design capacity. A substantial portion of total per barrel costs are determined by annual fixed costs divided by annual production volume. Consequently, effective per barrel production costs are quite sensitive to changes in plant capacity factor.

Table 3-4 illustrates this sensitive relationship between plant capacity factor and effective production cost. The first line of the table shows how capital costs per unit of product change as plant capacity factor is varied from the nominal assumption of 90 percent. (Recall that $24 per barrel was the nominal case capital cost based on $60,000 plant investment cost per daily barrel of production capacity, 0.13 capital charge rate, and 90 percent capacity factor.) The second line of the table shows total capital plus operation and maintenance (O&M) costs if one assumes that one-half of O&M costs of $15 per barrel are fixed costs which do not vary with production volume. Such fixed O&M costs magnify the effect of plant capacity factor on product costs. Although total costs are not shown in Table 3-4, they could be readily estimated by adding an assumed per barrel cost for feedstock coal from Table 3-3 to the second of Table 3-4. For example, if feedstock costs were $15 per barrel, total costs for a capacity factor of 75 percent would be $61 per barrel (46 + 15 = 61). The uncertainty regarding actual plant performance and capacity factors cannot be resolved via paper studies. Practial operating experience is required.[8]

Associated Infrastructure Costs. Estimates of the full costs of synthetic fuels production should include the costs for additional infrastructure development necessary to establish the production

Table 3-4. Effect of Varying Plant Capacity Factor on Per Unit Synthetic Fuel Costs

	Capacity Factor (percent of design capacity)					
	45%	60%	75%	90% (nominal value)	105%	120%
Capital cost per unit (1980$/bbl)	48	36	29	24	21	18
Capital plus O&M costs[a]	71	55	46	39	35	31

[a] Assuming one-half of O&M cost costs of $15 per barrel are fixed costs. Note that figures in this row *exclude* costs for coal feedstock.

facility, to operate it to produce products, and to deliver the products to existing markets or marketing systems. The nature and extent of these costs are quite situation specific. Costs vary with project location, the level of existing infrastructure to support the project, and the form of the project's product and its marketing system. Synthetic fuel projects located far from established population and marketing infrastructure may have significantly higher associated infrastructure costs than projects located near existing population centers and product marketing systems.

Two major sources of associated infrastructure costs are (1) community infrastructure developments required to support plant construction and operation and (2) transportation of synfuel products to established marketing systems. Example estimates for each of these categories of associated infrastructure costs are outlined below.

Population impacts and community infrastructure costs for synthetic fuels deployment were reviewed in the 1975 study by a federal interagency task force. The study estimated investment costs for community infrastructure at approximately $100 to $200 million (1980$) for a single coal-based synfuel plant located in a relatively sparsely populated area and requiring some development of new towns. Such costs are a substantial initial outlay which could cause serious cash flow problems for limited community budgets. Over the life of a production facility, however, such investment costs translate into per barrel costs of only about $1 per barrel.[9]

Estimates of the associated infrastructure costs for the new pipelines or marketing infrastructure required to transport synfuel products to market require assumption of particular project locations and situations. The proposed coal gasification project in North Dakota is a good example of these added costs. The special connection pipeline between the North Dakota synfuel plant and the existing interstate gas pipelines adds about $5 per barrel oil equivalent to effective product costs for that project.[10] For synfuel projects located near existing transportation, costs for product transportation to the marketing infrastructure may be substantially lower.

Total costs for new associated infrastructure to support synthetic fuel production and utilization could include other costs in addition to the two categories noted above. For example, costs may be incurred for new fuel distribution facilities to utilize a methanol synfuel product. Associated infrastructure costs are very case-specific and uncertain, ranging from very small (less than $1 per barrel) to moderately substantial (on the order of $5 to $10 per barrel). A reasonable range is about $1 to $5 per barrel.

Adding Components to Obtain a Range for Total Costs. A range of total synfuel production costs can be estimated by adding the ranges for individual components of per barrel costs. Table 3-5 dis-

Table 3-5. Ranges for Components of Total Coal-Based Synfuel Production Costs (1980$ per barrel of oil equivalent).

Cost Component	Approximate Range	Major Source of Uncertainty or Variation
Capital costs	15-40	Total plant investment cost Plant capacity factor Capital charge rate
Operation and maintenance costs	10-20	Process variation Lack of experience Total plant investment cost
Coal feedstock costs	7-15	Coal to product efficiency Coal cost/price
Associated infrastructure costs	1-5	Very situation specific
Range for total costs	33-80	

plays reasonable per barrel cost ranges for capital costs, O&M costs, coal feedstock costs, and associated infrastructure costs. Major sources of uncertainty or variation for each cost component are noted. The plausible range of total costs for coal-based synfuels is substantial, from about $30 to $80 per barrel oil equivalent in 1980 dollars. Assessments requiring estimates of synfuel production costs should explicitly recognize the inherent uncertainties and reflect an appropriate parametric range of synfuel cost outcomes.

Synfuel production costs are only the first component of calculations required to estimate net production subsidies (or savings). The calculation also requires estimates of the costs for the oil products displaced by synfuels. The next section reviews price trajectories for crude oil prices and related petroleum products.

Oil Price Trajectories and Petroleum Product Costs

The market value for synthetic fuel products will depend on the market prices for the petroleum-derived products which the synfuels displace. Petroleum product prices depend on the cost of crude petroleum used to produce a product as well as the product's form value. "Form value" refers to the different market values for petroleum products based on their differing chemical or physical performance characteristics and uses.

Oil Price Trajectories. Alternative scenarios for future oil price trajectories abound. The experience of the 1970s demonstrated the volatility of the international market and made carefully developed oil price projections seem embarassingly erroneous. An assessment of near-term glut and stable or declining oil prices, common in the 1977 context, was soon replaced by expectations for aggressive real increases in oil prices. The crisis tenor of 1979 and 1980 faded in 1981 as an excess of oil supply relative to demand softened oil prices. The only certain characteristic for future oil prices is that they are uncertain.

Rather than analyzing possible oil price futures, this study simply assumes a set of alternative long-run oil price trajectories. The primary set of oil price cases used in this analysis are the 1980 Department of Energy Policy and Fiscal Guidance (DOE-PFG) long-run oil price trends (Department of Energy 1980b). However, two other sets of price trajectories will be examined as sensitivity cases. The

Table 3-6. Constant Percentage Annual Growth Rate Oil Price Cases.

| | | \multicolumn{7}{c}{YEAR} | | | | | | |
Case	Annual Growth Rate	1980	1985	1990	1995	2000	2010	2020
High[a]	4.2%	30	37	45	56	68	100	100
Medium[b]	3.0%	30	35	40	47	54	73	80
Low[c]	1.8%	30	33	36	39	43	51	60

[a]Assumes maximum price level of $100/bbl in 1980$.
[b]Assumes maximum price level of $80/bbl in 1980$.
[c]Assumes maximum price level of $60/bbl in 1980$.

first set of alternative trajectories reflect percentage annual real growth in oil prices from a base level of $30 per barrel in 1980. Parameter assumptions and price schedules for these trajectories are displayed in Table 3-6.

A second set of oil price trajectories is also examined as a sensitivity illustration (Table 3-7). Sensitivity trajectory A views the present world oil market as disrupted and assumes that the long-run oil price trajectory will grow in real terms at about the real rate of interest from a 1980 undisrupted price level of $15 per barrel.[11] The real rate of interest is assumed to be 3 percent.

Sensitivity trajectory B assumes that real price increases occur in each decade but that the amount of real increase in each decade declines as prices increase in later decades. This assumed trajectory contrasts with the tendency of constant percentage growth rate cases to reflect increasingly large absolute price increases as higher price levels are reached. For this scenario, real prices start at $30 per barrel in 1980, increase by $20 per barrel through the 1980s, increase by $15 per barrel over the 1990s, $10 per barrel during the 2000-2010 decade, and $5 per barrel in the 2010-2020 decade. Prices are assumed to remain constant in real terms thereafter.

Sensitivity trajectory C assumes that oil prices remain constant in real terms at $35 per barrel (1980$) for the next several decades. Table 3-7 displays the price schedules for these three oil price trajectories, which are merely assumptions, not predictions. Such sensitivity assessments can help us appreciate the net costs or savings associated with the wide range of possible long-run oil price trajectories.

Table 3-7. Price Schedules for Example Sensitivity Oil Price Trajectories (average refiner acquisition cost for imported oil in 1980$ per barrel).

		Year							
Label	Summary Description	1980	1985	1990	1995	2000	2010	2020	2030
A	3% annual growth from $15/bbl in 1980	15	17	20	23	27	36	49	66
B	Declining increments each decade	30	40	50	58	65	75	80	80
C	Constant in real terms	35	35	35	35	35	35	35	35

Form Value Differential for Petroleum Products. Prices for crude oil translate into prices for the petroleum products derived from crude oil via the economics of the refining process and the marketplace. Different petroleum products—such as motor gasoline, distillate fuel oil, or residual fuel oil—follow different refining paths, have different characteristics and uses, and compete with different substitutes. Consequently various forms of petroleum products have different market values. Synthetic fuel products can substitute for a variety of different petroleum products. Accordingly, the market value or revenue available from the sale of synthetic fuel products depends on relative price levels prevailing for various forms of petroleum products.

These relative price levels depend on evolving factors, such as the price and characteristics of various crude oil supplies, the structure of refining capacity, and the nature of demand for various types of petroleum products. These factors, as well as others, result in gradually changing relative prices for various products. For example, the market value of residual fuel oil may decline as more coal-fired boilers are built, but increase as refineries invest in equipment to turn residual fuel oil into gasoline. Future relative product values may differ from those that exist today.

Nevertheless, the existing form value differentials provide an indication of the relative value of different forms of petroleum products. Table 3-8 displays average 1980 prices for various petroleum products compared to crude oil. Form value differentials are referenced to crude oil and expressed both in absolute 1980 dollars per barrel and percentages. Although future relative prices may change from those tabulated, the table shows that the market value of synthetic fuel products can vary significantly from a barrel of crude oil having the equivalent energy content. If synthetic fuel products directly substitute for petroleum-derived gasoline, favorable form value differentials could be about $10 to $20 per barrel. In contrast, a synfuel product that displaces primarily residual fuel oil may have a market value less than crude oil.

Technologies for the production of synthetic fuels from coal can yield a spectrum of products. Many of the technologies likely to be applied as part of an initial commercial production program, such as indirect liquefaction (Fischer-Tropsch synthesis, coal to methanol, etc.), produce products that may substitute for relatively high-valued petroleum products such as motor gasoline (Fluor 1979; Hottel and Howard 1971). A substantial form value premium relative to crude oil should be associated with such products. The market application and form value of other synfuel products is less clear. For example, the market value for methane produced from coal may depend on natural gas supply developments. The market value of a liquid boiler fuel derived from coal may depend on the availability and alternative uses for the residual fuel oil that would be displaced.

These form value differentials add another element of uncertainty to estimates of the net production subsidy costs. For this analysis, it is not necessary to develop detailed estimates of future form value differences for various synthetic fuel products as compared to crude oil. Form value variation will be incorporated within the general synfuel cost uncertainty. For example, a synthetic fuel product that costs $60 per barrel to produce, but that has a form value of $10 per barrel relative to crude oil, has an equivalent or effective cost of only $50 per barrel.

Estimates of Net Production Subsidy Costs

Net production subsidies are calculated by subtracting effective synfuel production costs from effective oil costs. "Effective" costs reflect values such as the oil import premium or the form value of

Table 3-8. Form Value Differentials for Petroleum Products.

Product	Price	Energy Content	Price	Form Value Difference Relative to Crude Oil[a]	
	(1980$/bbl)	(mmBtu/bbl)	(1980$/mmBtu)	Absolute (1980$/barrel)	Percentage
Crude oil	28.07	5.8	4.83	—	—
Leaded regular gasoline	38.22	5.3	7.21	14.0	+49%
Naphtha	35.61	5.4	6.59	10.5	+36%
Kerosene jet fuel	36.79	5.7	6.45	9.5	+34%
Diesel or distillate fuel oil	33.94	5.8	5.85	6.0	+21%
Residual fuel oil[b]	23.06	6.3	3.66	−7.0	−24%

[a]Calculated and expressed on a Btu basis assuming six million Btu per barrel; rounded to nearest $0.50.

[b]The *1980 Annual Report* presents data for a composite residual fuel oil only. However, most synfuel products would be relatively low in sulfur and thus would displace residual fuels with lower sulfur content. These generally have a higher market value than the composite value noted above. For example, EIA's *Monthly Energy Review* (May 1981) reports July 1980 crude oil prices of $28.73 (composite refiner acquisition cost), while residual fuel oil prices for that month were $24.89 for 0-0.3 percent sulfur, $23.44 for 0.3-1.0 percent sulfur, and $19.20 for greater than 1.0 percent sulfur content (weight percent).

Source: Derived from data presented in Energy Information Administration, April 1981, *1980 Annual Report to Congress, Volume Two: Data* (U.S. Department of Energy). 1980 prices from Table 40 (for crude oil) and Table 41 (for petroleum products). Petroleum product prices *exclude* taxes.

synfuel products. The comparison between synfuel production costs and oil costs may change over time as petroleum product prices evolve or as synfuel production costs change, but subsidy costs in any year equal the differential between effective synfuel costs and oil costs multiplied by production volume.

Opportunity costs for subsidies will be incurred when synfuel production costs exceed the cost for the oil products displaced. Generally, the amount of production subsidy should decline each year if oil prices steadily increase. (Synfuel costs are expressed on a levelized basis and are assumed not to increase over time due to increased coal prices and so forth.) The net present value for total production subsidies is calculated by adding together year-by-year quantities with appropriate discounting to reflect the time value of money for near-term subsidies (losses) that may be offset by out-year savings (profits). Calculations are carried out over an assumed thirty-year financial lifetime for a plant.

For assessments of the costs and benefits to the nation, the effects of inflation should be suppressed since inflation affects all investment opportunities. All costs and prices are compared in constant real 1980 dollars (1980$). As with the benefit estimates of Chapter 2, a real (inflation absent) discount rate of 5 percent is applied as the standard value. Estimates of net production subsidies are best presented in tabular format. Given the uncertainty for synfuel production costs, tables present net production subsidy costs for a parametric range of synfuel costs outcomes from $20 to $80 per barrel (1980$). Tables 3-9 through 3-13 present estimates of net production subsidy costs for alternative oil price trajectories, discount rates, and initial plant start-up dates.

The entries in these tables are net present value in 1980 expressed in billions of constant (real) 1980 dollars. Positive numbers indicate net savings as compared to oil; numbers in parentheses are negative and denote net subsidies. Tabulations represent net costs or savings for a single representative synfuel plant producing 0.1 Quad each year or about 50,000 barrels per day. Larger or smaller plants or production amounts would scale in direct proportion to the relative production quantities. For example, costs for three 50,000 barrel per day plants would be about three times the tabulated cost for a single plant, while costs for a 25,000 barrel per day plant would be one half of the tabulated value. Unless otherwise noted, the table calculations reflect a 5 percent discount rate and assume plants begin full production (come "on-line") in 1990.

Table 3-9 displays net subsidy costs for the DOE-PFG oil price trends. As expected, costs vary greatly with outcomes for long-run oil price trajectories and synfuel cost level. Production subsidies or savings on the order of billions of dollars net present value may be plausibly expected for a project. The values tabulated are quantitative estimates of the size of the insurance premium that may be paid by society for each initial production project.

Table 3-9. Production Subsidy Estimates for Department of Energy Policy and Fiscal Guidance Oil Price Trajectories (net present value for 0.1 quad production plant on-line in 1990; NPV of savings or (subsidy) in billions of 1980$ at 5 percent discount rate).

Synfuel Production Cost Level (1980$/bbl)	Policy and Fiscal Guidance (DOE-PFG) Trajectory		
	Low	Medium	High
20	2.5	4.3	6.0
30	0.8	2.5	4.2
40	(0.9)	0.8	2.5
50	(2.7)	(0.9)	0.8
60	(4.4)	(2.7)	(1.0)
70	(6.1)	(4.4)	(2.7)
80	(7.9)	(6.1)	(4.4)

Tables 3-10 and 3-11 show net effective subsidy costs for the two sets of alternative long-run oil price trends. Table 3-10 shows that initial synfuel production yields net savings rather than subsidy costs over the long-run if oil prices grow at more than about 3 percent per year and effective synfuel production costs are less than about $60 per barrel crude oil equivalent. Table 3-11 suggests that large subsidy costs are expected for the constant real growth from $15/bbl case or the constant $35/bbl scenario. Modest to substantial net savings relative to oil are projected for the oil price trajectory B, which reflects large price increments in early decades.

Table 3-12 shows the effect of varying discount rates on the expected net present value. Discount rates greater than the standard 5 percent reduce the discounted estimate for subsidy costs. However, since near-term subsidy years are weighted more heavily than out-year savings, higher discount rates also decrease the synfuel cost level for which net production subsidy costs are first estimated to be incurred. The zero discount rate case reflects subsidy requirements or production savings without discounting. Potential nominal dollar budgetary outlays for subsidies can be estimated from these undiscounted figures by assuming appropriate inflation factors.

Table 3-13 displays the effect of changing the date at which synthetic fuel plants begin initial production. Estimates are shown for plants coming on-line in 1987, 1994, and 2000. In general, changes

Table 3-10. Production Subsidy Estimates for Annual Percentage Growth Oil Price Paths (net present value for 0.1 quad production plant on-line in 1990; NPV of savings or (subsidy) in billions of 1980$ at 5 percent discount rate).

Synfuel Production Cost Level	Percentage Annual Growth Rate Case		
	Low (1.8%)	Medium (3.0%)	High (4.2%)
(1980$/bbl)			
20	4.2	6.4	8.9
30	2.5	4.7	7.2
40	0.8	2.9	5.5
50	(1.0)	1.2	3.7
60	(2.7)	(0.5)	2.0
70	(4.5)	(2.3)	0.3
80	(6.2)	(4.0)	(1.5)

Table 3-11. Production Subsidy Estimates for Sensitivity Case Oil Price Trajectories (net present value for 0.1 quad production plant on-line in 1990; NPV of savings or (subsidy) in billions of 1980$ at 5 percent discount rate).

Synthetic Fuel Cost Level	Oil Price Sensitivity Case		
	A 3% Annual Increase from $15/bbl	B Decreasing Decade Increments	C Constant $35/bbl
(1980$/bbl)			
20	1.6	7.7	2.6
30	(0.2)	5.9	0.9
40	(1.9)	4.2	(0.9)
50	(3.6)	2.5	(2.6)
60	(5.4)	0.8	(4.3)
70	(7.1)	(1.0)	(6.1)
80	(8.8)	(2.7)	(7.8)

Table 3-12. Subsidy with Varying Discount Rates for Department of Energy Policy and Fiscal Guidance Price Trajectories (net present value for 0.1 quad production plant on-line in 1990; NPV of savings or (subsidy) in billions of 1980$ at indicated discount rate).

Synfuel Cost Level (1980$/bbl)	Discount Rate and PFG Price Scenario					
	10%			0%		
	Low	Medium	High	Low	Medium	High
20	0.9	1.6	2.3	8.4	13.8	19.0
30	0.3	0.9	1.6	3.1	8.4	13.7
40	(0.5)	0.3	0.9	(2.2)	3.1	8.4
50	(1.1)	(0.5)	0.2	(7.6)	(2.2)	3.0
60	(1.8)	(1.1)	(0.5)	(12.9)	(7.6)	(2.3)
70	(2.5)	(1.8)	(1.1)	(18.3)	(12.9)	(7.7)
80	(3.2)	(2.5)	(1.8)	(23.6)	(18.3)	(13.0)

in the initial production date have only a small effect on net expected production subsidy. Given the extended financial life of the facilities (thirty years), outcomes for long-run oil price trajectories and synfuel production costs are more important determinants of expected subsidy or savings than the particular year in which synfuel plants begin production.

These estimates of net production subsidy costs for a single project illustrate the substantial range of social costs or savings that could be associated with an initial synfuel production program. Each production facility requires multi-billion dollar investments which, depending on plant performance or oil prices, could lead to subsidy costs on the order of several billion dollars net present value. With relatively high long-run oil prices or low synfuel costs, equally large savings as compared to purchase of imported oil are also possible. Near the middle of the ranges examined for synfuel costs and oil prices, the net present value of potential savings or subsidies is about two billion dollars for each representative synfuel project. An initial production program comprised of a number of facilities could have potential costs (or savings) that scale in direct proportion to the volume of production.

Table 3-13. Subsidy for Varying Plant Initial Production Date (net present value for 0.1 quad production plant; NPV of savings or (subsidy) in billions of 1980$ at 5 percent discount rate).

Synfuel Cost Level	Initial Production Date for DOE-PFG price trajectory								
	1987			1994			2000		
	Low	Medium	High	Low	Medium	High	Low	Medium	High
(1980$/bbl)									
20	2.7	4.7	6.7	2.3	3.7	5.1	1.9	2.9	4.0
30	0.7	2.7	4.7	0.9	2.3	3.7	0.8	1.9	2.9
40	(1.4)	0.7	2.7	(0.6)	0.9	2.3	(0.3)	0.8	1.9
50	(3.4)	(1.4)	0.7	(2.0)	(0.6)	0.8	(1.3)	(0.3)	0.8
60	(5.4)	(3.4)	(1.4)	(3.4)	(2.0)	(0.6)	(2.4)	(1.3)	(0.3)
70	(7.4)	(5.4)	(3.4)	(4.9)	(3.4)	(2.0)	(3.5)	(2.4)	(1.3)
80	(9.4)	(7.4)	(5.4)	(6.3)	(4.9)	(3.4)	(4.5)	(3.5)	(2.4)

FACTOR COST INFLATION

The estimates tabulated above define possible costs for an individual synfuel production project. However, a *program* for initial synthetic fuel production comprising many simultaneous projects could incur costs greater than the simple summation of the expected costs for individual plants. Increasing levels of aggregate demand for various inputs, or "factors" of production, required for the design and construction of synfuel projects could lead to price increases, scheduling difficulties, and other inefficiencies. This taxonomy applies the summary term "factor cost inflation" to these additional costs.

Sources of Factor Cost Inflation: A Qualitative Overview

Initial synfuel production efforts could cause demand for selected equipment items as well as for labor, engineering, and management services to exceed available supplies. One major engineering company has estimated that a large synthetic fuels program comprising twenty coal-based synfuel plants could require on the order of 15 to 25 percent of existing U.S. capacity for a number of major equipment items.[12] Another study of the architect, engineering, and construction (AE&C) industry indicates that approximately 10 to 15 percent of all chemical engineers projected to be available in 1985 would be required for an effort to build twenty coal liquids plants by 1990.[13] Simultaneous construction of many synthetic fuel facilities would also require significant numbers of highly skilled construction laborers, such as pipefitters and welders. Supplies of some skilled construction workers are presently short.

The result of this strain on existing supplies, capabilities, and infrastructure would be price increases or limitations on the availability of good and services. Such effects, though easy to predict qualitatively, are difficult to assess quantitatively. The amount of cost increases will depend on the level of demand required for a particular synfuels scenario as well as on market conditions. For example, the extent of price increases for a synfuels program will depend on the level of demand from related industries such as refinery or chemical plant construction. Price increases also will depend on how the suppliers of selected goods and services expand production capacity in response to higher demand, profits, and prices. In general, price increases for synfuels-related goods and services could result in substantial social costs for both the synthetic fuels program and other industries and economic activities that utilize the same inputs. (One should note, however, that for proper cost-benefit analysis, increases in the domestic price level are not necessarily real costs to the nation.[14])

The added social costs due to factor cost inflation stem from numerous sources. Some are obvious, others less so. The many possible effects of shortages or bottlenecks for construction inputs are illustrated by the 1973–74 experience of the chemical construction industry. During that period, high levels of construction demand led to supply bottlenecks and a period of hyperinflation in chemical plant construction costs. Costs increased at a rate of 10 to 15 percent above the general rate of inflation.[15] Costs experienced during this hyperinflation period included price increases, quality reductions, scheduling difficulties, reduced labor productivity, planning costs for deferred projects, and deferred investment opportunities. One study summarized the 1973–74 experience as follows:

> From the end of 1972 to the end of 1975 the cost of constructing a chemical process plant in the United States increased nearly 45% according to the DuPont index . . . . Many of the costs over that period will never be

fully recorded. Plant quality suffered as quality control on the part of equipment manufacturers and construction firms was relaxed to meet demand . . . . The DuPont index significantly understates real increases for plants under construction during the period as prices went from below list to over list plus surcharges . . . . Labor productivity fell dramatically for many projects as laborers waited for pacing equipment items to arrive . . . . The cost of planning projects that were subsequently cancelled (i.e., "false starts") and the lost industry expansion constitute major hidden costs of the period. (Merrow and Worthing 1979:15)

Some might consider such cost increases acceptable if focused only within a synthetic fuels effort. However, the effects of price increases for these primary production inputs, plus the induced shortages or "bottlenecks as a result of the synfuels program would have repercussions on the costs of all process plants and industrial construction which competes with synfuels plants for inputs" (Merrow and Worthing 1979:15). One engineering firm observed in its report to a U.S. Senate committee that "achievement of production levels on the order of 1 million barrels per day would require a national commitment for diverting our resources into this activity, deprive other sectors of the economy, and lead to sharply escalating costs" for related industries, equipment, and material (Cameron Engineers 1979:35). Such cross-industry impacts could increase substantially the effective costs to the nation of an initial synthetic fuel production effort.

Estimating Factor Cost Inflation by Analogy

Factor cost inflation is a key issue for the assessment and design of initial synfuel production programs. Unfortunately, several factors complicate the task of developing well-grounded estimates for factor cost inflation.

First, the possible shortages that could lead to price increases and project execution inefficiencies depend on the economy-wide demand *and* supply balances. Aggregate demand for potentially constrained materials or services depends on a number of market situations and scenarios in addition to the synthetic fuel deployment schedule. For example, variations in worldwide or domestic economic growth could markedly change net demand. Markets for construction labor and material in some regions could depend on the deployment scenario for the MX missile system or coal-slurry pipelines. The level of demand for chemical industry investment may depend on trade policies toward petrochemical imports.

Even if demand scenarios could be well-specified, estimates of existing supply capacity or supplier responses to particular demand or price projections is a difficult and subjective task. Most estimates of availability and price for projected synthetic fuel material, equipment, and service requirements rely on the experienced judgments of procurement personnel in major architect, engineering, and construction management firms.[16] Also, we have little experience regarding the nature of supplier responses to the high demand, price, and profit levels that go hand in hand with factor cost inflation. Consequently, estimates of the potential level of factor cost inflation that might be associated with any given level of initial synthetic fuel production are speculative at best.

Nevertheless, the experience of the 1973-74 hyperinflation period can be used to develop an estimate that illustrates the potential magnitude of cost inflation associated with rapid increases in construction costs or bottlenecks. Table 3-14 outlines a simple calculation that projects the rates of real construction costs increases (over and above general inflation) observed during the 1973-74 hyperinflation period onto a nominal six-year construction schedule for a coal-based synthetic fuel plant. If real construction cost increases of about 13 percent per year were realized, plant capital costs would increase about 39 percent over the construction period as compared to a case where construction costs increase at the same rate as general inflation. Capital-related costs (capital plus fixed O&M) comprise about two-thirds of total product costs. Total product costs, therefore, would increase about 20 to 25 percent if a synfuels program induced hyperinflation similar to that experienced in 1973-74.[17]

The Table 3-14 calculation demonstrates how such hyperinflation in construction costs can substantially increase synfuel production costs in real terms. Although it is difficult to assess the levels of construction cost increases that might be associated with any particular level of initial synfuel deployment, the levels of aggregate demand associated with the 1973-74 inflation provide an illustration. During the 1973-74 period, industrial engineering and construction contract volume peaked at about $40 billion (1980$). Since the average contract volume during the 1972-78 period was about $30 billion

Table 3-14. Example Calculation for Increased Synthetic Fuel Production Costs if Hyperinflation Similar to 1973-74 Period Is Experienced.

The first step is to estimate inflation in construction costs over and above general inflation:[a]

	1974	1973	Ratio (1974:1973)
A. Dupont index for total plant costs (1972 = 100):	135.4	109.8	1.233
B. Commerce Department GNP deflator (1972 = 100)	116.4	105.6	1.102
Difference			0.13 or 13%

(Line A ratio minus line B ratio equals real construction cost increase)

	Project Year					
Rate of Capital Expenditures:[b]	1	2	3	4	5	6
C. (% of total per year)	4%	10%	30%	34%	20%	2%
D. Real cost increase for capital expended in particular year	1.0	1.13	1.28	1.44	1.63	1.84
E. Expenditure weighted capital cost inflation for each year (Line C multiplied by Line D)	.04	.113	.384	.490	.326	.037

F. Total expenditure-rate weighted capital cost increase: 1.39 or a 39% increase (Sum of individual years in Line E)

G. Impact of capital cost increases on total product cost assuming about two-thirds of total product cost is capital-related[c]: 25% increase in total per barrel costs (0.65 × 39% = 25%)

[a] Cost indexes from E. I. DuPont de Nemours & Co. as reported in Edward W. Merrow and J. Christopher Worthing, 1979, "Possible Shortages in a Synthetic Fuels Mobilization," draft manuscript.

[b] Rate of capital expenditure schedule assumed by Department of Energy for coal liquids plants. From Department of Energy, "Energy Security Corporation Briefing Book," unpublish draft manuscript, p. F-11. Table 3.

[c] See Table 3-5, which suggests that over two-thirds of total product costs are capital-related. Note that capital-related O&M costs should be considered. Also, coal costs may increase if factor cost inflation drives up prices for coal mining equipment.

(Merrow and Worthing 1979; Cameron Engineers 1979) an excess demand of about $10 billion annually or 33 percent was associated with the 1973-74 hyperinflation period.

Contract volume for a synfuels production effort depends on the capital cost of the plants. Assuming our nominal capital cost of $60,000 per daily barrel of capacity, a deployment of one million barrels per day production capacity implies plant investment and architect, engineering, and construction contract volume on the order of $60 billion (1980$).[18] Even with contract volumes and construction projects spread over several years, the aggregate levels of activity required to achieve 1 MMBD synthetic fuels production are clearly significant relative to the level of demand which induced the 1973-74 hyperinflation.

The potential for construction cost inflation induced by synfuel deployment efforts is magnified if synfuels construction is not expected to crowd out competing non-synfuel projects. The costs of crowding out could be significant, since the need for new construction in capital-intensive industries could well increase in response to rising energy costs. For example, chemical companies might not undertake construction projects for energy conservation or for conversion from oil and gas to coal. In effect, overly aggressive synfuel deployment might displace investments for other potentially more cost-effective measures to reduce oil imports.

Although quantitative estimates are problematic, existing constraints for the engineering-industrial infrastructure required for synfuels deployment create the clear potential for factor cost inflation. One payoff of initial production experience is enhanced deployment capability—expanding the volume of production increments possible before cost-increasing constraints are encountered. Deployability benefits and added costs from factor cost inflation are two sides of the same coin.

Although estimates are difficult, factor cost inflation should become more serious as the size of an initial production effort expands. Moreover, factor cost inflation is likely to exhibit increasing marginal costs.[19] Twice the production volume may incur more than double the level of factor cost inflation. A small program may incur no factor cost inflation, and a moderate program only modest levels; however, a large crash program could well experience substantial additional costs, as observed in an assessment by Cameron Engineers:

> Because of the potential for hyperinflation, . . . costs could increase dramatically in a crash program. Building 20 plants could cost considerably more than twice as much as building ten plants. Any savings in design costs by building duplicate plants would be wiped out by cost increases. Plant construction costs during an all-out crash program are likely to increase by 50% or more. (Cameron Engineers 1979:27)

"WHETHER?" A SYNFUELS PROGRAM: CONCLUSIONS ABOUT COSTS AND COMPARISONS WITH BENEFITS

Two Qualitative Conclusions

This taxonomy and assessment suggests two major observations for the costs of initial synthetic fuel production efforts. First, each of the major components of program costs is substantially uncertain. Synfuel production costs could vary over a wide range. Moreover, even if synfuel costs were well-characterized, net production subsidies would remain uncertain because they depend on outcomes for future oil price trajectories. Costs for a program of many initial production projects also depend critically on the amount of factor cost inflation. These levels are very difficult to reliably estimate. Environmental and social costs depend on particular project choices and technology outcomes. Individual values and subjective judgments also affect assessments of intangible costs.

Second, factor cost inflation is likely to increase the average cost for each plant in larger initial production efforts. Per barrel production costs as well as environmental and social costs are likely to be greater for larger initial production efforts. The construction activity and associated impacts increase in direct proportion to program size. Moreover, rapid attainment of large production volumes may require special procedural treatment or substantive environmental provisions. Established political and institutional relationships may be increasingly strained as the scale and pace of initial production efforts increase.[20]

Figure 3-1 illustrates graphically the qualitative relationship between program size and program costs. For the first few plants, expected costs increase in direct proportion to program size. At some point, however, factor cost inflation makes total program costs increase faster than in direct proportion to program size. The point at which this occurs (point A in Figure 3-1) is uncertain. Also, the rate at which marginal program costs increase with increasing program size (the slope of section B of the curve in Figure 3-1) is uncertain. Despite the uncertainties about quantitative descriptions, the qualitative depiction of the program cost/program size relationship shown in Figure 3-1 is realistic and has important implications for the design of cost-effective initial production efforts.

Figure 3-1. Increasing Marginal Costs with Increasing Program Size Due to Factor Cost Inflation

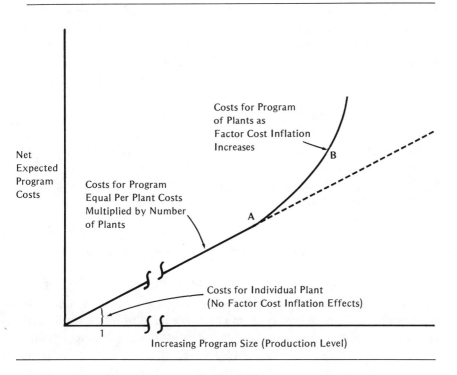

Example Estimates of Synfuel Program Costs

Although the many uncertainties make quantitative estimates problematic, appropriately caveated estimates can be useful. This chapter has developed estimates for three important elements of the costs of an initial synthetic fuel production effort.

First, a plausible range for synfuel production costs of $30 to $80 per barrel (1980$) has been described. Second, this range has been combined with an assumed range of oil price trajectories to develop parametric estimates of potential social costs or savings for synfuels as compared to oil. Subsidy costs for a representative plant could be several billion dollars (net present value).

Third, the 1973-74 experience of hyperinflation in chemical plant construction costs provided an analogy that suggests that a one million barrel per day initial production effort could plausibly experience factor cost inflation on the order of 25 percent or more.

Reflecting increasing marginal costs, a program half as large might experience less than half that level of factor cost inflation. For example, a 500,000 barrel per day program might experience 10 percent additional costs. A small initial production effort (say 250,000 barrels per day) might avoid factor cost inflation altogether. Such estimates are merely reasonable assumptions. The underlying phenomena are poorly understood. For example, if engineering and management capabilities are a key constraint, factor cost inflation may depend more on the number of major projects than on the total volume of production.[21]

Although these representative ranges and estimates for the costs of a synfuels program are plausible, the substantial uncertainties should be explictly recognized. Assessments should include sensitivity analyses for assumptions about key parameters.

The Paradox of Synfuel Costs, Savings, and Payoffs

Production of coal-based synfuels could yield either substantial net costs or savings relative to oil. If oil prices follow a low or moderate trajectory which is below synfuel costs, initial synfuel plants will incur substantial net costs for production subsidies. If oil prices follow a high trajectory exceeding synfuel costs, initial projects incur no costs and indeed yield savings relative to oil. (Note that if oil prices are certainly likely to be higher than synfuel costs, private firms would invest to realize these savings as profits.)

These outcomes present a paradox for the role of synfuels in our energy future. Synfuels become a profitable and attractive option in the unhappy world of the high oil prices we seek to avoid. However, in the more favorable oil price future of low to moderate long-run price trajectories, the net subsidy costs for initial production projects increase greatly. A favorable economic scenario for coal-based synfuels may well be an unfavorable energy future for the nation.

The strategic insurance framework for the benefits, costs, and purposes of initial production efforts helps clarify this paradox. For our broad national economic interests, we should hope that oil prices follow the low to moderate trajectory that implies substantial costs for production subsidies. However, we should be willing to accept these subsidy costs for some number of initial production projects as an insurance premium payment that will improve our national posture for responding to potential unhappy events such as rapid increases in the market price or security costs of imported oil. The subsidy costs necessary to foster initial production projects may be worthwhile, since the improved capabilities resulting from those projects provide valuable insurance payoffs should a surge deployment effort become necessary.

The Key Role of a Framework for Benefits and Costs

Part I of this study has shown that the potential benefits and costs of an initial synfuel production program are about the same size. The net balance depends on particular assumptions for uncertain parameters, outcomes, and scenarios. The answer to "Whether?" and "Why?" special investments in initial synfuel production experience are warranted is likely to be "Yes," primarily because of the improved strategic capabilities realized via initial practical experience with an important energy option.

The next question to be addressed is "How large?" an initial production program is appropriate. Part II examines this question. Developing an answer requires weighing benefits and costs for various program sizes. These comparisons should be made for many assumptions and scenarios and systematically integrated. Part II uses a "model" to make the many calculations and comparisons required; however, the parametric estimates of benefits and costs developed with straightforward arithmetic in Chapters 2 and 3 provide the basis for the model's comparisons across many scenarios.

The conceptual framework for benefits and costs that has been presented in Part I is critical to the entire study. The framework helps answer the important question of why special emphasis might be placed synfuels as compared to other long-run alternatives for reducing oil imports. Also, the benefit framework suggests why free market incentives alone may not adequately advance the nation's economic interests. Moreover, the information and insurance capability framework for the special benefits of initial synfuels production experience provides the underlying structure for the assessment in Part II of "How large?" an investment is appropriate. Finally, the taxonomy for benefits and costs helps focus important program purposes and issues for the discussion in Part III of "How to do better?" with initial synthetic fuel production efforts.

NOTES

1. See Merrow, Chapel, and Worthing (1979:95-108). Estimate of 20 to 25 percent for preliminary estimate drawn from Table 4.1. See also Merrow, Phillips, and Myers (1981) for statistical estimates of the expected amount of underestimation.
2. The annual dollar cost reported on the studies has been converted to percentage of plant facilities investment in order to facilitate comparisons with other estimates. "Plant facilities investment" (PFI) differs from the total plant capital investment costs by the cost for interest during construction, start-up expenses, working capital and initial catalyst/chemicals, royalties and land.
3. This assumption is reasonable within the range of accuracy sought here. For example, Oak Ridge National Laboratory (1981) estimates annual costs for catalysts and chemicals of about $13 million (1979$). These costs are more than offset by estimated by-product revenues from sale of ammonia ($7.6 million annually), phenols ($9.0 million annually), and excess electric power ($9.4 million annually). By-product credits might also be available for sulfur. Actual consumable O&M costs and by-product values vary substantially from process to process and with assumed marketing conditions for the by-products.
4. The calculation is straightforward: 6 percent times $60,000 for 365 days per year corrected for an on-stream factor of 90 percent yields $11/bbl (0.06 × 60,000/(365 × 0.9)). For 10 percent, the calculation is (0.10 × 60,000/(365 × 0.9)), or $18/bbl. The range in O&M costs could be larger. Lower capital costs are possible, as are lower capacity factors. Maintenance costs for large projects may not rise according to the percentages used here.
5. Price figures for Coal Delivered to Steam-Electric Utility Plants from *Monthly Energy Review*, February 1980, Department of Energy.
6. Coal to product conversion efficiencies vary noticeably among technologies and even with different product slates and operating modes for particular technologies. For example, Fluor estimates of coal to product efficiency for various indirect liquefaction product slates range from 49 percent for maximum liquids production, to 57 percent for co-product Lurgi methane and Mobile methanol to gasoline conversion (Fluor 1979:28).
7. The calculation is straightforward ($60,000 × 0.13) divided by (365 × 0.9) equals $23.74. The 0.9 factor reflects a 90 percent capacity factor.

8. The experience of the nuclear industry regarding capacity factors is a discouraging example. Availability and production rates for nuclear power plants have been substantially lower than many expectations held prior to substantial commercial operating experience (Joskow and Rozanski 1979). Empirical studies have shown that capacity factors for innovative chemical process plants are often substantially less than the 90 percent nominal assumption. Capacity factors are often especially low during initial years of operation as technology "bugs" are worked out. For a good discussion, see Merrow, Phillips, and Myers (1981).

9. Based on estimates derived from Volume II, of Synfuels Interagency Task Force (1975:D-17) entitled *Cost Benefit Analysis of Alternate Production Levels*. The estimate of $100 to $200 million is derived from the table by taking 40 percent of the community infrastructure costs for three high Btu gas from coal facilities. Forty percent was used because the table entry is for two plants located in remote western regions and one plant in a partially developed eastern area; development in western areas was assumed to require more additional investment for infrastructure than in eastern sites. The resulting per plant figure of $88 million (1975$) is updated to $123 million (1980$) using an inflation factor of 1.4. Investment costs of $123 million translate to a per barrel cost of about $0.97 per barrel, assuming a capital charge rate of 0.13 and a 90 percent capacity factor for a representative 50,000 barrel per day facility.

10. Estimated from data in Federal Energy Regulatory Commission (1979:17). Appendix A of this opinion notes transportation costs for new feeder pipeline from the plant to the existing interstate gas pipeline of $.82 per Mcf in 1979$. This translates into about $.90/Mcf in 1980$ or about $5 per barrel oil equivalent. This estimate may overstate costs expected, since significant economies of scale for the pipeline may be available as plant size increases from the one-quarter scale represented by the initial phase of the project to a larger plant. Nevertheless, this estimate is adequate for the rough approximating purposes intended.

11. Sensitivity Trajectory *A* reflects the economic theory of exhaustible natural resources. In simplest terms, economic theory maintains that the rents for a depletable resource should increase at about the real rate of interest. For example, see Hotelling (1931) and Solow (1974). The $15 (1980$) starting price reflects the assumption that in the early 1980s the oil market has not yet stabilized from a period of disruption. Sensitivity Trajectory *A* assumes that in the 1978 period the oil market had stabilized at about $15 per barrel in 1980$.

12. Estimated in Fluor (1979:38). Numerical estimates drawn from Figure 6-11. Estimate reflects "percentage of existing U.S. vendor capacity required to support the total 1,500,000 B/D Synfuels Program." The percentage in any particular year was not noted.

13. Estimate derived from Dinneen, Merrow, and Mooz (1981). Estimate of about 23 percent for a scenario of 1.6 MMBD of coal synthetics by 1990 was scaled down about 15 percent for a 1.0 MMBD effort (23% ÷ 1.6 = 14.4%).
14. There is an important distinction between price increases and real costs to society. Shortages may induce temporary price increases above the marginal cost of production or "rents" for the providers of goods and services in short supply. If payments take place entirely within a given definition for society—such as U.S. society—such rents represent transfers of wealth within society rather than real social costs. Consumers sacrifice wealth to producers. Consumers are worse off, but producer's welfare increases. Total social welfare (consumer plus producer welfare) remains the same. Thus rents can represent primarily the substitution of producer well-being for that of consumers. Only a portion of the price increases, induced by the added demand of a synfuels program, necessarily translates into increases in real social costs. However, increases in market prices for some inputs may cause adjustments, such as deferred consumption or investment. Services or goods may be purchased abroad. Such adjustments represent real social costs. The distribution of costs stemming from factor price inflation between transfers within society and real social costs is a complicated, but potentially important, assessment. As a simplifying convention, this study will assume that price increases ultimately will reflect real social costs rather than simply transfers within society. (For more background on the concept of economic rents, see Mansfield (1970).)
15. Derived from Dupont data for process plant prices as reported in Merrow and Worthing (1979:8). General inflation rate based on Commerce Department GNP Deflator.
16. One of the more systematic assessments of availability of resources for energy development is a study done for the U.S. government by the Research and Engineering Division of Bechtel Corporation (Gallagher, et al. 1976). These assessments rely primarily on demand estimates generated by the Bechtel Energy Supply Planning Model combined with supply availability judgments of experienced Bechtel commodity advisers, engineering managers, and administrators.
17. Analogous estimates for alternative rates of increase in real construction costs (over and above general inflation) of 5, 10, or 20 percent yield corresponding product cost increases of about 7, 15, and 32 percent, respectively.
18. A similar assessment of design and construction contract volume is reported by Cameron Engineers (1979:37-38). Their analysis shows that a crash synthetics program would lead to contract volume levels associated with the 1973-74 hyperinflation.
19. The expectation that factor cost inflation is likely to increase on the margin for larger programs reflects an economic model for the phenomena involving

COSTS FOR INITIAL SYNFUEL PRODUCTION EFFORTS 111

inelastic factor supply curves. Factor price inflation occurs as the demand levels induced by the synfuel production efforts encounter relatively inelastic portions of factor supply curves. Where supply curves are inelastic, doubling demand more than doubles price. If supply elasticity equals one, doubling demand levels doubles price. Since price increases for factors of production increase marginally with larger programs, total product costs and levels of program factor cost inflation should also reflect increasing marginal costs.

20. For a particularly effective discussion of how political and socioeconomic costs grow as the scale of initial production efforts increases, see Seidman (1980).
21. Assessments of the architect, engineering, and construction industry suggest that ten to twenty major commercial synfuels projects may strain existing capabilities, thus introducing added costs. For example, Cameron Engineers (1979:36) notes that only twenty-one large design and construction firms have contracted for work near the amount required by even a small synfuels project ($400 million). Merrow and Worthing (1979) suggest that only about eight projects could be undertaken immediately with firms clearly capable of managing a project of the scale of a synfuels plant. Dinneen, Merrow, and Mooz (1981) characterize significant limits on the existing synfuel capabilities of the AE&C industry.

11 "HOW LARGE?" A SYNFUELS PROGRAM
The Decision Analysis Assessment

Part I has shown that the benefits and costs of an initial synthetic fuel production effort are about the same. The net comparison depends on particular assumptions for uncertain parameters, outcomes, and events.

Benefits exceed costs primarily when unfortunate events—such as a sudden increase in the market price or security costs of imported oil—motivate a surge deployment effort. Initial synfuel production experience acts as an "insurance policy" for such "insurance event" scenarios. The costs or "insurance premiums" are paid in the form of subsidies for initial production projects. The "insurance payoffs" are improved capabilities with the synfuels option that result from practical production experience. On balance, Part I has shown that the purchase of some insurance is likely to be worthwhile.

Insurance can be purchased in varying forms and amounts. Given the enormous investments and potential costs, a key question is "How much?" insurance or "How large?" an initial production effort is appropriate. This question does not have obvious answers. Net benefits and costs vary significantly with different outcomes for key parameters such as synfuel production costs and long-run oil price trends. For example, when surge deployment events occur, benefits grow large as strategic capabilities are used, and costs for production subsidies decline with increased costs for imported oil.

Different sources of insurance benefits and costs also vary with program size. Basic information about synfuel production costs can be realized with one or two initial production projects. The majority of learning benefits can be realized via a small to modest initial production effort; however, given a critical mass effect for infrastructure development, large programs may more effectively enhance deployment capability. Yet larger programs may also incur added near-term costs from factor cost inflation.

The "How large?" question becomes further complicated when the opportunity for a phased approach to the insurance purchase is recognized. Both the cost of the insurance and the likelihood of using the synfuels option depend on the uncertain cost level for synfuel production. Initial production experience will clarify this key uncertainty. If synfuel costs turn out to be happily low or moderate, subsequent production expansion decisions—either to build infrastructure capability or install production capacity that is economic compared to oil—can be aggressive. Conversely, if synfuel costs are high, or outlooks for oil prices more stable and moderate, second phase decisions for production or insurance capability can be appropriately moderated. Both the insurance costs and likely value of practical production experience should be clarified in the first phase of an initial production effort.

Part II presents a systematic quantitative assessment of the decision concerning "How large?" an initial synfuels production effort should be. The analysis is based on the framework for benefits, costs, and surge deployment events characterized in Part I. The formal economics and operations research technique called "decision analysis" is applied to the complex assessments because it offers two major advantages for assessing the synfuels insurance investment.

First, a decision analysis structure allows the inherent uncertainties for many key parameters to be separately considered and characterized, yet systematically integrated. As shown in Chapters 2 and 3, relative costs and benefits depend on combinations of outcomes for several key parameters (such as synfuel costs, oil price trends, the amount of learning and deployability growth, and so forth). One person could not easily consider, compare, and integrate the many uncertain factors and scenarios which bear on the appropriate program size. Using a decision analysis approach, however, a person's assessment about the likelihood for each of a range of possible outcomes can be specified for important parameters, addressing one uncertain parameter at a time.

Moreover, this advantage is not obtained at the cost of ignoring the full range and combinations of outcomes that should be considered in a systematic assessment of costs and benefits. For example, an extreme combination of outcomes such as low synfuel costs and high oil prices could result in extremely large benefits; a contrasting combination would result in high costs. An averaging approach, where outcomes for key uncertain parameters are assumed at representative values in the middle of their uncertain range, may not properly weigh into the decision assessment the possibility or impact of potential high benefits or large cost scenarios. The decision analysis technique, however, allows extreme scenarios, such as a surge deployment event, to be explicitly evaluated and weighed, in proportion to their likelihood of occurring, into an integrated assessment of expected benefits and costs.

A second major advantage of decision analysis is its capacity for incorporating phased or sequential decision points. Decisions on production capacity need not be made now for all time. An initial phase can be followed, as warranted, by either aggressive expansion, modest expansion, or no expanded production at all. Information from initial production will clarify appropriate future choices. Given the time required to develop infrastructure and capability, however, the range and cost of options at future decision points will depend on the character and extent of prior production experience. A decision analysis structure can reflect both the value of improved information about key uncertainties and the benefits of an improved and expanded range of options for future decisions.

Presentation of the decision analysis assessment is structured into four chapters. Chapter 4 describes the general structure of the decision analysis framework and model. Each element of the analysis structure is described on a general basis before numerical assumptions are specified for a base case example. It is important to recognize that the model can be used as a tool to examine the implications of alternative assumptions for key parameters and probabilities. The base case assumptions have no special significance.

Chapter 5 describes how the decision analysis model works. Quantitative results have limited value unless they are understandable and believed by the persons whose decisions might be assisted by the analysis. The decision analysis model performs straightforward calculations and comparisons across the many scenarios that determine the expected benefits and costs of initial synfuels production efforts. Chapter 5 endeavors to de-mystify the "black-box" model calculations by relating the model's operation to the parametric estimates developed with simple arithmetic in Chapters 2 and 3. The expected or probability-weighted outcomes of contrasting extreme scenarios are used to illustrate the straightforward accounting function of the decision analysis model. Chapters 4 and 5 attempt to develop the understanding that will help the model results be recognized as well-grounded and plausible, even if they are counter-intuitive.

Chapter 6 presents results of the decision analysis assessment. Base case model outputs are reviewed in some detail to illustrate the range of information available from the decision analysis results. However, since results necessarily depend on input probabilities and assumptions, the bulk of Chapter 6 reviews a number of sensitivity cases. These cases illustrate the implications for the "How large?" question of changing specifications for such key assumptions as the probability of price jump events, the relative importance of learning and infrastructure development benefits, and the amount of factor cost inflation. A final set of sensitivity cases addresses the important question of whether private market incentives alone are likely to lead to the level of initial production experience and synfuels capability that seems appropriate for the nation as a whole.

Chapter 7 concludes Part II by summarizing the results of the base case and numerous sensitivity cases with three general observations on the "How large?" question. First, bigger is not necessarily proportionately better. There is a general trend toward declining marginal benefits for increasing program size. A large program can be worse than more modest efforts or no initial production program at all.

Second, preferred program size is especially dependent on a few key parameters and effects. High synfuel production costs or low long-run oil price trends greatly reduce the value of larger programs. The likelihood of surge deployment efforts is a strong determinant of the appropriate program size. The added near-term costs from factor cost inflation very quickly offset any added benefits from large programs. The importance of infrastructure development has strong implications for appropriate program size.

And third, a moderate size program is a good choice for a wide range of views for key parameters and probabilities. The analysis shows that this choice is very often the best decision and, even when it is not technically the best, it is nearly as good as the best.

The discussion in Part II is lengthy. It attempts to present carefully not only the results of the decision analysis assessment (Chapter 6), but also how the model is structured (Chapter 4) and operates (Chapter 5). Readers are encouraged to focus their attention on those parts of the presentation that are of particular interest to them. For example, a reader interested primarily in analysis results may find it useful to skim or skip Chapters 4 and 5, referring back to those chapters as necessary to clarify questions about the modeling that underlies a particular result. Other readers may be more interested in the structuring of a decision analysis model for an insurance problem having the size and complexity of the synfuels situation. Those readers may wish to emphasize Chapter 4. Part II's length stems from the desire to allow interested readers who are not extensively familiar with decision analysis techniques to understand the model structure, operation, and results.

The results of Part II's decision analysis assessment of the "How large?" question may seem plausible to some, counter-intuitive to others, and obvious to others. The analysis structure underlying these results, however, responds to many of the concerns and arguments of both synfuels advocates and those less enthusiastic about special programs to promote initial synfuels production. Some concerns of synfuels advocates addressed in the analysis conclude:

- Uncertain and potentially high long-run oil price trends;
- Potential oil price jump events;
- Long-run security costs of oil imports and the potential for geopolitical events, which could suddenly increase those costs;
- Technology improvement learning effects; and
- An industry infrastructure effect whereby options for future production levels depend on the levels of prior production experience.

The cost-benefit analysis structure also responds to many concerns of those less enthusiastic about a major subsidy program for initial synfuels production. Some of their concerns include:

- Uncertain and potentially high synfuels costs;
- Added costs incurred for large ("crash") initial production efforts;
- Possibility for phased decisions on synfuel production levels that can respond to evolving information about key uncertainties; and
- The basis for a proper role for government.

Ultimately, the answer to the "How large?" question depends on individual views for key parameters and probabilities. There can be no general single answer, but Part II shows that systematic analysis can shed useful light on this important issue concerning appropriate allocation of our limited resources. Moreover, the analysis helps characterize the expectations about our energy future that make different levels of initial synthetic fuel production seem worthwhile.

4 GENERAL STRUCTURE OF SYNFUELS DECISION ANALYSIS FRAMEWORK AND MODEL

This chapter presents the general structure of the decision analysis framework and model for assessing the benefits and costs of initial synthetic fuel production efforts. It begins by providing a basic introduction and simple example of the application of the decision analysis technique to a simple insurance problem. The example establishes some terminology and a basis for understanding the subsequent description of the structuring for the much more complicated synfuels insurance problem.

STRUCTURING A DECISION ANALYSIS FRAMEWORK FOR A SIMPLE INSURANCE PROBLEM

Decision analysis is a well-established operations research technique for assessing appropriate choices in the face of uncertainty and extreme outcomes. Insurance situations inherently involve uncertain events, such as the occurrence of an accident, and uncertain but potentially extreme outcomes. The value of insurance and the willingness of a buyer to purchase various amounts of it depend on the likelihood of events and the costs of the insurance compared to the payoff from the insurance should an insured event occur. For example, if I live next to a stream that floods on average once every ten years, I am more interested in flood insurance than my neighbor who lives ten feet up the hill and whose house is usually unaffected by the rising waters. My probability for an insurance event is higher than my neighbor's.

The cost of an event or the payoff from insurance coverage also affects my willingness to purchase insurance. If, when a flood event occurs, my sunken living room with wall-to-wall carpeting is devastated, I am more interested in insurance than if the flood only affects my concrete-floored basement recreation room. Even my neighbor slightly up the hill may be interested in flood insurance if a once-in-fifty-years flood would devastate the Persian carpets and antiques in his sunken living room.

Decision analysis is a technique for systematically laying out the combinations of a decision and various uncertain outcomes that define a particular scenario. After each scenario is specified, the benefits and costs for that combination of outcomes can be calculated. The benefits in one scenario can be compared to the costs in others. The appropriate decision depends on the relative likelihood of various scenarios. These relative likelihoods depend on the probability for the scenario as defined by the combined probabilities for the various uncertain outcomes that make up a particular scenario.

A simple example can illustrate the structural elements and function of a decision analysis "tree." Suppose I am assessing the question of whether to buy flood insurance. To keep things simple, let me assess two choices—no insurance and full insurance with no deductible. I will represent my decision point as a square box with branches coming out of it. Each branch corresponds to one of my two alternative insurance choices.

Following my insurance-purchase decision, uncertain outcomes occur. These uncertain outcomes are represented by a circle with different branches coming out which correspond to each alternative outcome. A different probability is associated with each outcome.

Suppose there are two key uncertainties for my flood insurance problem. First, does a flood occur or not? And second, if a flood occurs, will I be the home to move the $3,500 worth of stereo, television, and furniture out of the flooding basement rec-room? If I am home to move these items, the cost of the flood is only $1,500. If not, the cost if $1,500 plus $3,500 to replace ruined furnishings. Figure 4-1 displays the structure of this flood insurance decision.

Figure 4-1. Decision Analysis Structuring of Simple Flood Insurance Problem

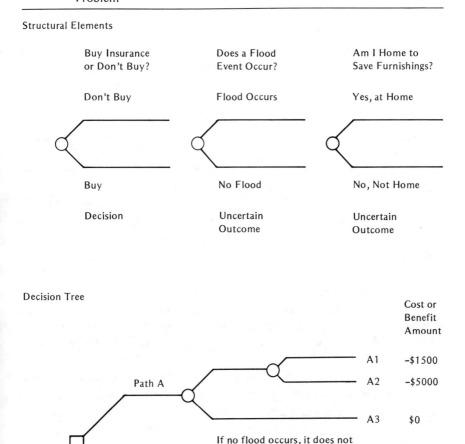

The top half of Figure 4-1 displays the structural elements of this decision analysis framework for assessing a flood insurance purchase. Alternative decisions are represented by "branches" from a square box or decision "node." Different uncertain outcomes are represented by branches from a circle or "chance node."

The bottom half of Figure 4-1 illustrates the source of the term decision "tree." When alternative decisions and uncertain outcomes are combined in sequence, the result is an expanding network of branches from various nodes that resembles the spreading branches of a horizontal tree. A "path" through the decision tree is simply a specific combination of choices proceeding sequentially through the tree from the base of the tree (initial decision) to the tip (last uncertain outcome). For example, one path through the tree (path $A1$ in Figure 4-1) is represented by the combination of no insurance (a decision), a flood occurs (an uncertain outcome), and presence at home to move furnishings (an uncertain outcome). Another path or scenario (B in Figure 4-1) is the decision to purchase insurance. If insurance is purchased, uncertain outcomes for a flood event or being home are not relevant since the dollar cost to the homeowner does not depend on either uncertain outcome.

Each path leads to a specific level of cost or benefits which is calculated at the tip of the tree after each scenario has been fully specified. For example, Path $A1$ from Figure 4-1 leads to no insurance payment but losses from the flood of $1,500. (If one was not home to move furnishings (Path $A2$) losses would be $5,000.) Path B leads to payment of the insurance premium, but no flood losses are ever incurred. If insurance costs $275 per year, then the cost of Choice B is always $275. Preference between no insurance (Choice A) and full insurance (Choice B) depends on the relative likelihood of each path.

The expected costs for Choice A depend on the probabilities for a flood occurring and the probability that someone will be home to move furnishings should a flood occur. Assuming a one in ten chance for a flood (10 percent probability) and a three out of four (75 percent) chance that someone will be home during a flood, the combined probability for the $1,500 loss associated with Scenario $A1$ is 7.5 percent or about one in thirteen. Path $A2$ has greater costs, since no one is home to move furnishings. This scenario occurs with a probability of 2.5 percent or about a one in forty chance. In Scenario $A3$, the flood does not occur, so neither flood losses nor insurance premium payments are incurred. This favorable scenario occurs with 90 percent probability.

The expected value of the "no insurance" choice is the probability-weighted average of the three different paths possible for the "no insurance" choice at the decision node. This calculation shows expected flood losses (or the benefits from full flood insurance) are about $237. (The calculation is: [.075 × $1,500 = 112.5] + [.025 × $5,000 = 125] + [.9 × 0 = 0] = $237.50.) If the full flood insurance costs more than $237, I would be better off without the insurance taking my chances on losses should a flood occur.

However, the desirability of the insurance changes with varying probabilities. If there is only a 50 percent chance that I will be home if a flood occurs (because I spend the early spring in Florida), the expected losses from the "no insurance" choice are $325 (0.05 × $1,500 + 0.05 × $5,000 + 0.90 × $0 = $325). I am better off with the flood insurance premium of $275 than the expected flood losses of $325.

This simple flood insurance example illustrates the importance of both structure and assumptions for the appropriate choices characterized by a decision analysis approach. An insurance decision obviously depends on the probabilities for events. However, the inclusion of the uncertain outcome as to whether I was home or not is an important structural element of the decision problem that, depending on probability assignments, can affect appropriate choices.

This chapter describes the decision analysis framework and model used to assess the "How large?" question for initial synthetic fuel production efforts. Both the general structure of the decision tree and particular parameter and probability assumptions are important determinants of the results from the decision analysis calculations. Accordingly, the general structure of each decision or uncertain outcome node is described first. Following that, parameter and probability assumptions are specified for a sample base case.

GENERAL STRUCTURE OF THE SYNFUELS DECISION ANALYSIS

The primary decision addressed in this analysis is the amount of initial production experience that provides an acceptable level of insurance capabilities for potential surge deployment events. Figure 4-2 illustrates three major elements of uncertainty for this decision. The appropriate amount of insurance to "buy" depends on the cost or premium for the insurance, the value or payoff of the insurance, and the probability of an insured event that activates realization of insurance payoffs. A decision analysis model provides a tool with which to evaluate the purchase of the insurance capabilities fostered by an initial synfuels production program. Basic elements of the model are described below.

Figure 4-2. Major Elements of the Synfuels Insurance Problem

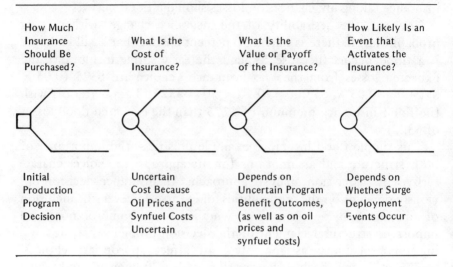

Decade-by-Decade Decision Structure

The choices to undergo decision analysis should be carefully defined. Although a variety of program strategies could be evaluated as separate program decisions, this analysis defines the primary program decision as the size of the initial production program in the 1980s. "Size" is defined as the coal-based synthetic fuel production capacity to be achieved in the 1980s decade.

Decisions on synthetic fuel production levels are not made once for all time. Phased production expansion decisions are built into the analysis structure with decisions on levels of production expansion being made in 1990 for plants to begin production in 2000 and in year 2000. (Plants are assumed to begin production ten years after the decision to build them is made.) The range of production increments possible for these 1990 and 2000 production expansion decisions depends on previous production levels and the strength of the infrastructure development effect. Thus, the decision analysis is structured into three decision epochs. Each epoch is one decade long, with decisions on synfuel production expansion occurring at the beginning of each decade. As Figure 4-3 illustrates, events occur, or uncertain outcomes are determined, between decision points.

At each of these decision points (the square boxes in Figure 4-3), the decision analysis model compares the expected value (net bene-

GENERAL STRUCTURE OF FRAMEWORK AND MODEL 125

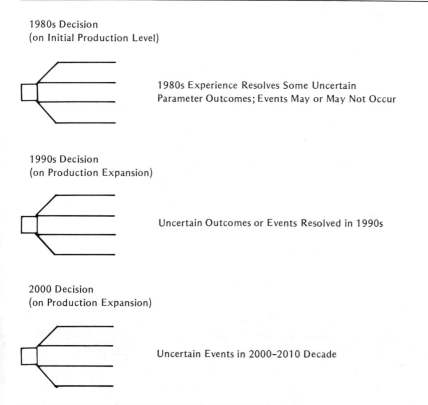

Figure 4-3. Three Decade Structuring for Synfuels Decision Analysis

fits) for each of four alternative choices that represent a range of production expansion decisions. The production expansion decision that provides the greatest expected net benefits is chosen.

Improved information about uncertain parameters and scenario outcomes can affect choices at production expansion decisions in 1990 and 2000.[1] For example, information about whether synfuel costs are high or low can improve production expansion decisions. If the synfuel cost outcome determined from 1980s experience is high, the 1990s production expansion decision will avoid incurring the subsidy costs of deploying synthetic fuels unless oil price expectations are also high. Without such information on synfuel costs, 1990s deployment may have been greater and additional costs incurred. On the other hand, if synfuel costs prove to be low, 1990 decisions can more confidently favor aggressive deployment.

Figure 4-4. Representation of Uncertain Outcome Node and Branches.

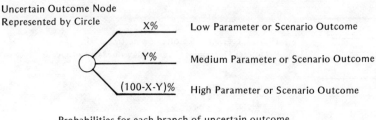

Probabilities for each branch of uncertain outcome
(total probability for all branches must equal 100%)

Key Uncertain Outcomes

Decision analysis allows each of a set of uncertain parameter, scenario, or event outcomes to be explicitly addressed via uncertain outcome or "chance" nodes. Each of these chance nodes characterizes a range of outcomes for an uncertain parameter (or scenario) and the probabilities ascribed to each particular outcome. For example, the basic level for synfuel production costs could range from $35 to $65 per barrel crude oil equivalent. A technological optimist would assert higher probabilities for the lower cost outcomes than someone more skeptical of synfuel technology costs and performance. Each different outcome for an uncertain parameter, trajectory, or event defines an additional scenario or path through the decision analysis tree. The different possible outcomes may be represented as alternative branches from an uncertain outcome node, as shown in Figure 4-4.

Each branch of the uncertain outcome node occurs with its own probability. The sum of the probabilities for all branches equals 100 percent. For example, if one were reasonably sure of the low parameter outcome, one might ascribe a 70 percent probability to the low outcome branch, a 20 percent probability for the medium outcome branch, and a 10 percent probability for the high outcome branch.

Each branch in a decision tree implies an additional set of scen-

GENERAL STRUCTURE OF FRAMEWORK AND MODEL 127

arios to be evaluated. In order to limit computational scope,[2] the range of uncertainty for a parameter or event is represented by a limited number of alternative outcomes (branches), usually two or three branches for each node. Furthermore, the computational size of the decision tree is kept manageable by limiting the structure to include uncertain outcome nodes only for those uncertain parameters or event outcomes that are key determinants of net program benefits and costs.

Seven uncertain outcomes are important to the assessment of costs and benefits for an initial synthetic fuels production effort. Long-run *oil price* (*OP*) trends and basic *synfuel* production *costs* (*SC*) are primary determinants of the subsidy costs for initial synfuels production as well as of the long-run benefits of improved capabilities with the synfuels option. *Program benefit outcomes* (*PBO*) for cost reducing *learning* (*PBO-L*) and *enhanced deployability* (*PBO-D*) describe the extent to which capabilities are improved. The net payoffs if a surge deployment event occurs depend on how much learning or infrastructure development has been realized from initial production experience. *Surge deployment events* (*SDE*) reflect the possibilities for sudden, discontinuous changes in the market price (*SDE-P*) or security costs (*SDE-S*) of imported oil. Finally, the maximum mid-term market role for synfuels might be limited by factors unrelated to infrastructure for synfuel deployment. Such a *synfuel production limit* (*SPL*) would make large deployment capabilities unusable and would reduce the volume of synfuel production providing benefits.

These seven key uncertain outcome nodes need not occur in each decision epoch. The decade-by-decade structure allows some uncertain events to occur or not occur in each decade. Moreover, some uncertain parameter outcomes may be resolved as experience is acquired in a particular decade.

Figure 4-5 summarizes the decision node and chance nodes that occur in the 1980s decade. Six chance nodes are included in the 1980s decade except when the "no program" option is chosen. Without any initial production experience, uncertainties about the basic synfuel cost level and the strength of program benefit outcomes remain unresolved. As a result, the decision analysis structure for the "no program" choice has only three uncertain outcome nodes in the 1980s decade.

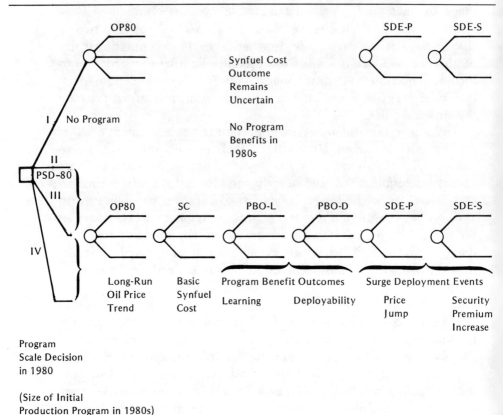

Figure 4-5. Structure of Decision Tree for 1980s Decade

MODELING FOR PARTICULAR ELEMENTS OF THE DECISION TREE

This section describes the modeling of various elements of the synfuels decision analysis tree. The full structure of the decision tree is built gradually by describing the relationships and outcomes represented by each of the decision and uncertain outcomes nodes. The general structure of each node is described first, then particular numerical values for parameters and probabilities for a base case analysis are specified.[3]

It must be emphasized at the outset that the numbers specified for the "base case" represent only one possible set of views for the various parameters and probabilities. The base case has no special importance. It merely reflects one person's (this author's) view of representative parameter and probability assignments. The decision analysis model is a tool. It can be used to assess the implications for appropriate program size of any particular view of the relevant parameters and likelihoods. Analysis outputs necessarily depend on input assumptions. Accordingly, most of the results presented in Chapter 6 are sensitivity cases that examine the implications for the "How large?" question of changing assumptions for key input descriptions and parameter values. The following sections describe each node of the decision analysis structure proceeding sequentially through the tree.

Program Scale Decision in 1980 (PSD-80) and Factor Cost Inflation

General Structure. This initial decision node represents the primary decision to be evaluated with the decision analysis tool—the best program choice for an initial synthetic fuel production effort. Conceivably, a wide range of alternative program strategies could be evaluated. For example, a small initial program with expanded technology research, development, and demonstration (RD&D) could be compared with a large commercial production program with limited RD&D, and so forth. However, this analysis considers only four alternative programs. Each program option is defined primarily by its size—the amount of initial production sought in the 1980s.

Initial program scale decisions affect some, but not all, of the subsequent uncertain outcomes and future decision options. For example, although long-run oil price trends or the likelihood for surge deployment events are unlikely to be affected by the level of initial synfuel production chosen in 1980, expectations for program benefit outcomes, such as the amount of learning or infrastructure development, may depend on the size or type of program undertaken.[4]

Factor Cost Inflation. Chapter 3 described the potential for additional costs for larger deployment efforts due to increases in the effective costs for some factors of production required for design, construction, and operation of a synfuel plant. Such factor cost inflation raises the cost level for synfuel projects built simultaneously as part of an initial production program or surge deployment effort. Factor cost inflation is modeled as a percentage increase in cost level realized for a group of plants deployed in a particular decade.[5] The increase is over and above whatever synfuel production cost might have applied in the absence of factor cost inflation. For example, if the basic synfuel cost outcome was $50 per barrel, 20 percent factor cost inflation would raise the synfuel production costs from $50 to $60 per barrel. As noted in Chapter 3, if factor cost inflation occurs, it will be greater for larger deployment efforts and decrease toward zero for smaller deployment efforts.

Factor cost inflation can significantly reduce net program benefits. For example, the $10 per barrel synfuel cost increase from the example above would decrease the net benefits of an initial production program comprised of 20 plants by over $34 billion net present value (NPV).[6]

The expected amounts of factor cost inflation must be specified for various amounts of initial production. Although factor cost inflation has a critical impact on the net costs and benefits, the levels of factor cost inflation expected for various size deployment efforts are quite uncertain. Accordingly, the sensitivity experiments reported in Chapter 6 include a large group of cases that examine alternative specifications for factor cost inflation. These sensitivity cases help characterize what one has to expect about the level of factor cost inflation to favor larger or smaller initial production programs.

Base Case Specifications. Four alternative program choices are examined at the 1980 decision point. A "no program" option is assumed to result in no significant experience with production of coal-based synthetic fuels in the 1980s. The basic synfuel cost level remains uncertain, and no learning or infrastructure development takes place. This "no program" option is a reference case against which the expected benefits and costs of other programs are measured. The three other program choices evaluated are a small program of 250,000 barrels per day in the 1980s, a moderate program of 500,000 barrels per day, and a large program of 1,000,000 barrels per day production capacity of coal-based synfuels by the end of the 1980s decade. If initial commercial projects are nominal size facilities of about 50,000 barrels per day, the 4 production levels translate into programs involving 0, 5, 10, and 20 projects.

Table 4-1. Base Case Specifications for Program Size and Factor Cost Inflation.

Program Label	Production Level in 1980s (millions of barrels per day oil equivalent)	Number of Nominal Production Facilities[a]	Amount of Additional Cost Due To Factor Cost Inflation
I: No program	0	0	0%
II: Small program	0.25	5	0%
III: Moderate program	0.50	10	+10%
IV: Large program	1.0	20	+25%

[a]A nominal or full-size facility is assumed to produce 50,000 barrels per day oil equivalent.

Only larger initial production efforts are likely to strain existing deployment capability causing factor cost inflation. The base case assumes that the second largest initial production level, 10 plants in the 1980s (Program III), experiences 10 percent factor cost inflation. The largest program, 1 million barrels per day in the 1980s (Program IV), is assumed to incur 25 percent factor cost inflation. These values reflect the view that factor cost inflation is likely to increase on the margin with increasing program scale. The base case specifications also reflect the quantitative illustration of the potential magnitude of factor cost inflation presented in Chapter 3. The base case assumes that no factor cost inflation occurs for the small program (0.25 MMBD or Program II). Table 4-1 summarizes these base case specifications for program size and factor cost inflation.

Long-Run Oil Price Trends (OP-80, OP-90)

General Structure. These chance nodes reflect alternative uncertain outcomes for the long-run trend of world oil prices. The oil price *trend* is a long-run smoothed trajectory reflecting evolution in energy markets according to alternative views of underlying geological, technological, and economic situations. The oil price trend concept specifically avoids attempting to reflect very transient oil price changes associated with various developments that affect oil supplies or price for only several months or a few years, such as politically motivated production cutbacks (embargoes) or temporary production losses due to technical accidents (pipeline failure). The possibility for jagged, discontinuous shifts in market price, such as the stepwise price increases experienced twice in the 1970s, will be reflected by the uncertain outcome for price jump events, to be described later.

There are numerous alternative views for future long-run oil price trajectories. The decision analysis model can reflect any set of price trajectories. Given the inherent uncertainties, the analysis should incorporate a set of trajectories that covers an appropriately broad range. For example, even though oil price increases were aggressive during the 1970s, future oil price trajectories could be much more moderate. A steady path of aggressively rising real prices is also possible. Even if one thinks a particular trajectory is much more likely to occur than others, balanced analysis should not totally neglect the possibility of substantially different oil price paths. Probability assignments can reflect an individual's view of the world regarding the relative likelihood of alternative long-run trends.

Oil Price Trend After 1990. Information about the likely long-run trend for oil prices can clarify appropriate decisions. If oil prices are trending high, synfuels deployment is more likely to be a good decision than if they are likely to follow a low trajectory. The general long-run trend for oil prices may become increasingly clear as experience accumulates through the 1980s, and, accordingly, an uncertain outcome for the long-run trend is included in the 1980s decade. Since it is unlikely that 1990 decisions on synfuel production expansion can be based on certain knowledge of the long-run oil price trajectory, an additional uncertain outcome node for long-run oil prices has been included in the decision tree after the 1990 production expansion decision. This chance node in the 1990s decade (*OP*-90) allows oil prices to shift from whatever long-run trend was followed in the 1980s to a different long-run trend.[7]

Base Case Assumptions. The range of long-run trajectories assumed for this analysis are the three 1980 Department of Energy Policy and Fiscal Guidance (DOE-PFG) long-run crude oil price trajectories. These three long-run planning trajectories reflect an appropriately wide range of plausible outcomes for smooth, gradually evolving

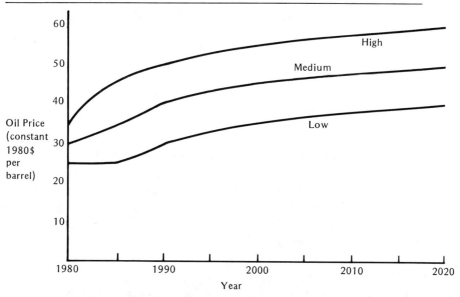

Figure 4-6. 1980 Department of Energy Policy and Fiscal Guidance Long-Run Planning Oil Price Trojectories (Refiner Acquisition Cost of Imported Oil In 1980$ per barrel)

world oil prices.[8] Price schedules for these trajectories were presented in Chapter 2, Table 2-5. Figure 4-6 displays the three DOE-PFG trajectories graphically.

Individual expectations for the likelihood of various long-run oil price trends vary greatly. The base case will assume a 25 percent probability for the low and high oil price trajectories and a 50 percent probability for the medium case.

Probabilities for a shift in oil price trends after the 1990 decision from one long-run trend to another also must be specified. The base case assumes a preference for continuing the oil price trend followed during the 1980s. This view of oil price trends and uncertainties is implemented by assuming a 60 percent probability for staying in the trend determined by the 1980s oil price outcome. Shifts to other long-run trends are possible, but shifts from one trend to the next higher or lower trend are more likely than shifts from low to high or high to low. Table 4-2 summarizes both the base case specifications for probabilities of alternative long-run oil price trends and also shifts from one trend to another.

Table 4-2. Base Case Probabilities for Oil Price Trends.

DOE Policy and Fiscal Guidance Price Case	Probability for 1980s Outcome	Probabilities for 1990s and Beyond Given 1980s Outcome of		
		Low	Medium	High
Low	25%	60%	20%	10%
Medium	50%	30%	60%	30%
High	25%	10%	20%	60%

Basic Synfuel Production Cost Outcome (SC)

General Structure. Chapter 3 presented reasons for the substantial uncertainty that remains regarding the basic level for synfuel production costs. We do not really know whether synfuel production costs are "forty-ish" or "sixty-ish." The answer has important implications for when and if the synfuels option should be extensively utilized. Practical production experience from initial production projects should eliminate many sources of uncertainty about production costs. Subsequent deployment decisions could be based on firm knowledge of the basic costs required to produce synthetic fuels from coal.

The adjective "basic" is important. It refers to the general cost level for synfuel production before accounting for learning, site-specific cost advantages or disadvantages, and other variations specific to particular technologies or situations. Knowledge of the basic cost level defines whether synfuel production costs are about $65 or about $35, not whether they are $62.40 for technology A and $66.15 for technology B. Just as for the oil prices to which these synfuel production costs are compared, measurement is in constant 1980 dollars per barrel of crude oil equivalent. The form value of various synfuel product types should be reflected in the crude oil equivalent cost.

Uncertainties about the basic level of synfuel production costs will be resolved in the 1980s decade except for the "no program" alternative. Since the "no program" case assumes that no significant commercial production experience with coal-based synthetic fuels will be obtained in the 1980s decade,[9] the uncertainty about

Table 4-3. Base Case Assumptions and Probabilities for Synfuel Cost Outcomes.

Basic Synfuel Cost Level	Probability
(1980$/bbl crude oil equivalent)	
$35	20%
$50	45%
$65	35%

synfuel production cost remains, at least for the 1990s decision. Thus, the decision tree structure for the "no program" choice does not include an uncertain outcome node for synfuel production costs in the 1980s. (Resolution of uncertainty about synfuel costs for the "no program" choice does not occur until practical production experience is first obtained in the 1990s or later.)

Base Case Assumptions. The range for synfuel production costs developed in Chapter 3 extended from $30 to $80 per barrel crude oil equivalent. Base case specifications assume three alternative synfuel cost outcomes of $35, $50, and $65 per barrel. Slightly higher probabilities will be applied to the middle and higher cost outcomes. Table 4-3 summarizes. The expected value of synfuel costs is the probability-weighted average, or $52.25 per barrel.

Program Benefit Outcome Nodes (*PBO-L, PBO-D*)

Advanced learning and enhanced deployment capability are key sources of insurance payoffs from initial synthetic fuel production efforts. However, the magnitude of these program benefits are uncertain. We have limited understanding about the slope of the learning curve or the strength of the linkage between production experience and expanding future deployment capability.

Moreover, the likelihood of various amounts of learning or deployability increase may depend on the size and design of the initial production efforts undertaken. Figure 4-7 illustrates the contrasting relationships of program size to learning benefits as compared to deployability benefits. Figure 4-7a shows that the majority of learning benefits can be realized with a small to moderate size program.

Figure 4-7. Contrasting Relationship of Program Benefits to Program Size

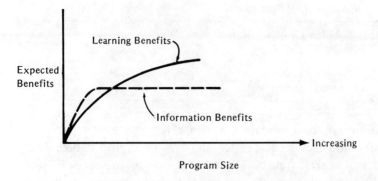

Figure 4-7(a). Learning and Information Benefits

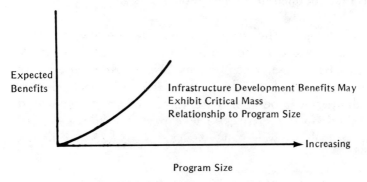

Figure 4-7(b). Deployability Benefits

In contrast, deployability benefits may increase substantially with larger initial production efforts (Figure 4-7b). The decision analysis model should reflect both the uncertainty about the level of program benefits as well as the contrasting relationships between expected benefits and program size.

Accordingly, the decision analysis structure includes two uncertain outcome nodes for program benefit outcomes (*PBO*). The uncertain outcome for amount of learning is reflected in a program benefit outcome for learning (*PBO-L*) node. Uncertain program benefit outcomes for deployability (*PBO-D*) are determined at a separate chance node. Uncertainty about the strength of learning and infrastructure development effects should be resolved via practical production experience in the 1980s.[10]

Program Benefit Outcomes for Learning (PBO-L)

General Structure. A range of uncertain outcomes for the amount of learning expected from initial production projects is reflected by a three branch program benefit outcome for learning (*PBO-L*) node. The learning outcome (*PBO-L*) is measured as the percentage cost reduction from the basic synfuel cost level that is achieved as a result of both selection learning (which technologies to use) and technology improvement learning (better versions). This percentage cost reduction is realized for each decade of experience with synthetic fuels production. If production continues in a second decade, the same amount of learning observed in the initial decade is assumed to apply again.

Expectations for the amount of learning may vary with the nature and extent of the experience from initial production efforts. Larger programs should have slightly higher probabilities for greater amounts of learning. However, learning with synthetic fuels is largely a sequential phenomenon where increasing the number of simultaneous projects may not lead to proportionate increases in the expected amount of learning. (See discussion in Chapters 2 and 8.) Consequently expectations for PBO-L outcomes do not increase as strongly as a function of the size of an initial production program. Figure 4-7a illustrated these declining marginal returns for learning as the size of an initial production program increases.

Base Case Assumptions. The appropriate range for the amount of learning from practical production experience is the subject of considerable uncertainty and debate.[11] Reflecting this this uncertainty, the base case considers a broad range for the amount of learning extending from 2 percent to 15 percent reduction in synfuel production costs following each decade of sequential production experience.

Although larger initial production efforts should yield somewhat more learning, learning gains will not increase in direct proportion to program size. The qualitative relationship between amounts of learning and program size depicted in Figure 4-7 is achieved by applying increasing probabilities for higher learning outcomes as program size increases. However, as shown in Table 4-4, base case probabilities for higher learning outcomes do not reflect a strong relationship between program size and the expected amount of learning. This structure for probability assignments causes learning

Table 4-4. Base Case Probabilities and Values for Learning Outcomes (probabilities for various amounts of learning).

Program Level	Size of Learning Cost Reduction (percent reduction from basic synfuel cost)		
	2%	8%	15%
I	(No learning until synfuel production experience obtained)		
II	35%	45%	20%
III	30%	45%	25%
IV	25%	50%	25%

Note: Probabilities add horizontally to equal 100 percent.

benefits to exhibit declining marginal returns to increasing program size.

Program Benefit Outcomes for Deployability (PBO-D)

General Structure. Program benefit outcomes for deployability (*PBO-D*) may exhibit a much stronger relationship to program size. The larger an initial production effort, the more extensively the engineering and industrial infrastructure required to deploy synthetic fuel production capacity is exercised and developed. Future maximum deployment capabilities may increase substantially more for larger initial production efforts than for smaller program. As suggested by Figure 4-7b, a critical mass effect could cause deployability benefits to reflect increasing marginal returns to program scale. Accordingly, the chance node for *PBO-D* reflects outcomes for the strength of the infrastructure growth factor that depend on program scale in the 1980s.

The simple formula for expanding deployment capability introduced in Chapter 2 is structured into the decision analysis model. Maximum deployment increments for the 1990 and 2000 production expansion decisions depend on three parameters: a base deployment capability (*BDC*), some deployability increase factor (*DIF*), and the level of production in the previous period. Two of these parameters are dependent on the level of initial production efforts in the 1980s.

GENERAL STRUCTURE OF FRAMEWORK AND MODEL 139

The base deployment capability (*BDC*) reflects the existing capacity of the U.S. economy to design, construct, and operate coal-based synthetic fuels production facilities irrespective of the levels of previous production. *BDC* is measured as the number of representative facilities that can be constructed in each decade. Deployment capabilities beyond this basic capability of the existing infrastructure must be obtained via practical production experience.

The quantitative linkage between production experience and developing infrastructure for enhanced future deployment capability is represented by the deployability increase factor (*DIF*). If building one facility in the 1980s allows two *additional* facilities (over and above the base deployment capability) to be built in the next decade, then the DIF factor equals two. However, if building one facility in the 1980s yields the capability to build an additional one-half facility, then the DIF factor is 0.5. If previous production experience does not increase future deployment capability at all, then DIF is zero. (In such a case, initial synthetic fuel production experience does not yield deployability benefits, since the future maximum deployment capability is not linked to the levels of prior production experience.)

Maximum possible deployment capability in any decade equals the base deployment capability (BDC) plus the level of production in the previous decade multiplied by the deployability increase factor (DIF). Thus, for a BDC of 20 projects per decade and a DIF of 1.5, a 10 plant initial production program would allow a maximum production expansion in the 1990s of 35 plants (20 + 1.5(10) = 35), or about 1.75 million barrels per day production capacity.

A key parameter in this description of increasing maximum deployment capability is the deployability increase factor (DIF). Unfortunately, the quantitative linkage between production experience and increased future deployment capability is not well understood. Accordingly, a range of DIF values are examined as different branch outcomes at the *PBO-D* node. The probabilities for each branch will depend on the scale of initial production program efforts. In general, as the size of initial production efforts increases, higher probabilities should be applied to the stronger infrastructure development effects or higher DIF outcomes.

This structure for modeling deployment capabilities allows future deployment capability to grow in two ways. First, future maximum deployment capability increases in direct proportion to the size of initial production efforts, irrespective of particular outcomes for the DIF parameter. Second, larger initial production programs will have higher probabilities for the larger DIF outcomes. As a result, larger initial production programs reflect increasing marginal returns to scale for deployability benefits as illustrated by Figure 4–7b.

Base Case Assumptions. Because the phenomena that underlie growing deployment capability are not well understood, the appropriate range of outcomes for the DIF or infrastructure growth factor is uncertain. The assessment in Chapter 9 suggests that engineering and project management capabilities for a complex multibillion dollar petrochemical project are likely to be a key deployment constraint. If so, one initial production project might make about one additional project possible in the next period. The DIF parameter would be about one. (See discussion in Chapter 9 for qualitative background for this view.) Accordingly, the base case assumes a set of outcomes for the uncertain infrastructure growth ranging from a weak linkage (0.3) to a relatively strong linkage (1.4). Base case probability assumptions assign much higher probabilities to stronger linkages for larger initial production programs. Table 4–5 summarizes base case specifications for DIF outcomes and associated probabilities for various initial production program sizes. As shown by the column at the far right of the table, the values specified increase markedly the number of additional plants possible for larger initial production efforts.

Surge Deployment Events (*SDE-P, SDE-S*)

Rapid shifts in the market price or social costs of imported oil could result in situations where aggressive deployment of synfuels plants could yield substantial benefits to the nation. Accordingly, the decision analysis structure for assessing the insurance costs and payoffs of initial synfuel production efforts explicitly includes uncertain outcome nodes that reflect possible sudden increases in the value of synthetic fuels production. Two separate types of surge deployment events (SDE) are considered: oil price jump events (*SDE-P*) and security events (*SDE-S*).

General Structure for Oil Price Jump Events (SDE-P). A major jump in the market price of oil is denoted as a "surge deployment

GENERAL STRUCTURE OF FRAMEWORK AND MODEL 141

Table 4-5. Base Case Expectations for Magnitude and Probability of Deployability Increase Outcomes (probabilities for various DIF outcomes).

Description of Initial Production Program		Outcome for Deployability Increase Factor			Expected Increase in Maximum 1990 Deployment Capability[a] (MMBD)
Label	Number of Nominal Plants	0.3	0.7	1.4	
I	0	(Base deployment capability only)[b]			0
II	5	45%	35%	20%	0.17
III	10	25%	45%	30%	0.41
IV	20	15%	45%	40%	0.92

[a] "Expected increase" refers to the probability-weighted average for the amount of additional deployment capability. This is calculated by multiplying the various probabilities times the various DIF outcomes for each program. The expected DIF is then multiplied by the volume of initial production to determine the increase in production possible due to enhanced deployment capability. Maximum production increments will depend on the assumed value for the basic deployment capability. If the base case assumption for BDC of twenty nominal projects is used, total production increments in 1990 equal 1.0 MMBD plus the values tabulated in the far right hand column above.

[b] For the "no program" choice 1990 maximum deployment capability equals basic deployment capability (BDC) only. The base case assumption for BDC is 1.0 MMBD.

Note: Probabilities add horizontally to 100 percent.

event—type P (for price)," or "*SDE-P*". The potential for rapid step-wise changes in the market price for oil was twice demonstrated in the 1970s. A price jump event (*SDE-P*) represents a substantial deviation from some long-run, smoothed, evolutionary trend for oil prices. (This long-run trend is the uncertain outcome of the *OP-80/OP-90* chance nodes.)

If synthetic fuels production capacity to be usefully deployed in response to price jump events, the deviation from the long-run trend must be moderately long lasting relative to the time required to deploy synthetic fuels production capacity. A temporary price spike of less than several years duration is not necessarily a development that should motivate surge deployment. Synthetic fuels are a long lead-time and long lifecycle option for which deployment decisions should be assessed relative to long-term trends and relatively long-lasting events.

Accordingly, a price jump event (*SDE-P*) is modeled as a doubling of oil prices at the end of a decade (1989, 1999, 2009) followed by a linear decline over fifteen years back to some long-run underlying trend. Such *SDE-P* events can occur in each decade, 1980–1990, 1990–2000, and 2000–2010. If an event occurs in a particular decade, it will increase substantially the value of synthetic fuel production decided upon and built in the following decade.[12] Oil price jump events also increase the benefits (or decrease the subsidy costs) for synfuel plants built before the event occurs.

General Structure for Security Events. The full cost to the nation of imported oil may not be fully reflected by its market price alone. These additional social costs are represented by an oil import premium. A portion of the oil import premium reflects the imputed costs for defense, foreign policy, and other social policies associated with the vulnerability of oil supplies to disruption or political manipulation. Geopolitical developments could result in marked changes in the security premium associated with oil imports. For example, a change in government or policy in Saudi Arabia could change our expectations for the probability, purpose, and consequences of oil supply disruptions or the political concessions extracted in return for stable oil supplies. One *SDE-S* event might be Soviet control of substantial portions of Persian Gulf production and the concomitant threat that such control would create for the industrialized democracies of Europe, Japan, and the United States. Even if the market price for oil remained relatively constant, potential changes in the security premium associated with oil imports could markedly increase the value of synthetic fuels production as well as other long-run measures for reducing oil imports.

A separate chance node represents the possibility for such a security-related surge deployment effort—*SDE-S*, S for security. An *SDE-S* event is modeled as a substantial increase in the long-run security premium for oil imports. Such an increase in the security premium could be associated with changed geopolitical and strategic assessments of the situation for the international oil market. A *SDE-S* event is *not* a relatively incidental short-term even, such as an embargo or threat of embargo over a passing issue. Rather, it represents a fundamental change in the security risks and costs associated with the role of world-traded oil for our economic, political, and strategic interests.[13]

If a security event occurs, the oil import premium is assumed to increase from a nominal level of about $5 per barrel (1980$) to a substantially higher value of $20 per barrel. These increased social costs due to security concerns will increase substantially the value of all import reduction measures, including synthetic fuels production. As with a price jump event, deployment decisions following an *SDE-S* event will be more aggressive. Even expectations for the possibility of a security event (*SDE-S*) act to increase the expected value of synthetic fuel production.

Both the price jump (*SDE-P*) and security-related (*SDE-S*) surge deployment events either occur or do not occur each decade. The possibility for surge deployment events is represented by two branch chance nodes in each decade. The occurrence (or not) of a price jump in one decade does not affect the likelihood for price jump events in later decades. However, once a security event occurs and increases the long-run oil import premium, the higher security premium is assumed to continue over the timeframe for the analysis.

Base Case Probabilities for Surge Deployment Events. Probabilities for surge deployment events are necessarily subjective. The experience of the 1970s and the volatility of the Persian Gulf region suggests that fairly high probabilities for a major oil price jump are not inappropriately pessimistic. Accordingly, the base case specifications assume a fifty-fifty chance for a major fifteen-year oil price jump event (*SDE-P*) occuring in each of the three decades considered for this analysis.

Political events in the Middle East—such as the turmoil in Iran, the Soviet invasion of Afghanistan, the Iran-Iraq war, and continuing Arab-Israeli tensions—have intensified concern for geopolitical developments that might increase the security costs of oil supplies from that region of the world. Again, subjective assumptions are required. The base case assumes that in each decade there is a 10 percent probability for a major long-term security event.

Table 4–6 displays base case probability assumptions for each type of event. The entries in Table 4–6 reflect the probability that a price jump event occurs in each decade. Although the model could examine different probabilities for events in various decades, the base case assumes, for simplicity, uniform probabilities for surge deployment events in each decade.

Table 4-6. Probabilities for Surge Deployment Events in Each Decade.

	1980s Decade	1990s Decade	2000 Decade
Price jump event (SDE-P)	50%	50%	50%
Increase in security Premium[a] (SDE-S)	10%	10%	10%

[a]Probabilities indicated assume that no security event has occurred in a previous decade. Once an *SDE-S* event occurs, the higher security premium applies for the remainder of the analysis timeframe.

1990s Production Expansion Decision (*PED*-90)

Following the decision and outcomes for the 1980s decade, a decision point on production expansion for the 1990s decade is reached. Because of the infrastructure development effect, the maximum possible production expansion depends on the 1980s' production decision, particular outcomes for the deployability increase factor determined at the PBO-D node, and the base deployment capability. The 1990 decision is modeled as whether to expand at the maximum rate possible or at some fraction of the maximum rate. Four different rates of expansion are allowed: no expansion, maximum expansion, one-third of maximum, and two-thirds of maximum.

The amount of factor cost inflation associated with various expansion decisions depends on the extent to which maximum deployment capabilities are utilized. For this analysis, maximum possible deployment increments in the 1990s or 2000s (2000–2010) decades are assumed to experience the same levels of factor cost inflation as the largest program choice for the 1980s. The second largest deployment increment (two-thirds of maximum possible) experiences factor cost inflation of the same magnitude as the second largest 1980s program. The third largest production expansion in the 1990s or 2000s incurs the amount of factor cost inflation specified for the third largest initial production program.[14]

At a decision node, the decision analysis model selects the amount of production expansion with the greatest expected net benefits. This ability of the decision analysis model to evaluate and select the "best" rate of expansion (of the four choices) makes the decision analysis model particularly appropriate for evaluating a phased program structure. Phased decisions can reflect the value of information obtained about key uncertainties resolved prior to the decision point. If the decision analysis model were not "smart" at decision nodes, the losses from bad decisions would be averaged (with probability weighting) with the benefits of the "best" decision. Such averaging would inappropriately reduce estimates for expected costs and benefits. Where a phased program structure and sequential decisions are possible, the decision analysis model can properly reflect the value of information from the first phase, which may improve second phase choices.

Following the 1990s production expansion decision, additional chance node outcomes occur in the 1990s. The uncertain outcome node for the post-1990 oil price trend has been described already. Uncertain outcome nodes for surge deployment events also occur again. The 1990s decade includes one additional uncertain outcome that is not clarified in the 1980s—the synfuel production limit.

Synfuel Production Limit (SPL)

General Structure. Many energy projections have suggested that at energy price levels similar to the production cost of coal-based synthetic fuels, U.S. oil imports may be completely eliminated by a combination of domestic energy supply and demand responses other than coal-based synfuels. For example, one 1980 DOE projection for "high oil prices and balanced high supply," projects zero oil imports in the year 2000. (Department of Energy 1980b: 35) For a number of other projections, conservation and enhanced energy production from sources other than coal-based synfuels result in oil import levels less than half of the 8 MMBD that the U.S. was importing toward the end of the 1970s.[15]

However, the maximum mid-term market role for coal-based synfuels may be limited even if maximum synfuel deployment capabilities are very large. Coal production constraints could arise as high annual production levels encounter logistical problems and congestion. Also, coal or fossil fuel "moratoria"—due to concerns for the climatological effects of increasing carbon dioxide levels—might limit coal production levels and the maximum amounts of coal-based synfuel production.

The synfuel production limit is a summary representation of the various uncertain developments for energy supply and demand, which could act to limit the maximum role for coal-based synthetic fuels as a component of the energy supply mix in the early part of the twenty-first century. By the end of the 1990s, accumulated experience with energy supply and demand responses may clarify uncertainties regarding the mid-term maximum role for the synfuel option. Accordingly, the 1990s decade includes a chance node (SPL) reflecting three alternative outcomes for the synfuel production limit. Aggregate production levels that can be reached with the year 2000 production expansion decision are not allowed to exceed this synfuel production limit, which is expressed in millions of barrels per day total coal-based synfuel production capacity.

The synfuel production limit acts to reduce the quantity of synfuel production over which learning benefits are obtained. It also reduces the value of infrastructure development, since the maximum deployment capability in many scenarios would cause cumulative production levels to exceed the synfuel production limit.

Base Case Assumptions. Because of the many uncertain outcomes for energy supply and demand that are summarized by the synfuel production limit parameter, base case specifications should reflect a wide range of outcomes. A low value of 2.5 million barrels per day (MMBD) reflects optimistic results for the contributions of domestic supply and demand alternatives other than coal-based synthetic fuels. A high value of 8 million barrels per day represents a role for coal-based synthetic fuel production as a domestic supply source and alternative to imported oil that is essentially unconstrained by contributions to the energy supply mix from other sources. A high outcome might also reflect U.S. export of coal-based synfuels to other oil importing nations. A mid-range outcome of 4.5 MMBD is simply an intermediate assumption roughly congruent with many projections of oil import levels in the early part of the next century.

Table 4-7 displays base case probability specifications for the three synfuel production limit outcomes. The expected (probability-weighted) value is five million barrels per day maximum production level for coal-based synthetic fuels through 2010.

Other Outcomes

1990s Outcomes for No Program Choice in 1980s. Three additional uncertain outcome nodes occur in the 1990s if no synfuels produc-

Table 4-7. Base Case Specifications for Synfuel Production Limit Outcomes.

Label	Limit Through 2010 (millions of barrels per day oil equivalent)	Probability
Low	2.5	25%
Mid	4.5	45%
High	8.0	30%

tion takes place during the 1980s decade. In such a "no program" case, basic synfuel cost levels and the rate of learning and infrastructure development will remain uncertain until initial practical production experience is acquired. If the 1990 decision results in some synfuels production, chance nodes for the basic synfuel cost level and program benefit outcomes occur in the 1990s decade. The probability and parameter assignments for these nodes are the same as those applied to programs II, III, and IV for the 1980s. Basically, for the case reflecting no program in the 1980s, the synfuel cost and program benefit nodes are shifted from the 1980s decade to the 1990s decade.[16]

2000 Production Expansion Decision. The year 2000 production expansion decision is structured like the 1990 decision. Maximum production expansion possibilities are linked to previous production decisions and outcomes. As for the 1990 decision, there are four alternative levels of expansion for the year 2000 including choosing none, all, or some part (one-third or two-thirds) of the maximum production expansion possible. Also, since the outcome for the synfuel production limit is determined in the 1990s (before the year 2000 decision), the maximum production increment is limited to an amount that prevents total production from exceeding the synfuel production limit.

Outcomes After the Year 2000 Decision. Following the year 2000 production expansion decision, two chance nodes recur for surge deployment events in the 2000-2010 decade. The benefits from surge deployment effort plants deployed in response to those events are not included in the decision analysis benefit calculations, since the analysis excludes the benefits of synfuel plants deployed after the 2000-2010 decade.[17] However, the possibility that oil price

Figure 4-8. Full Decision Tree Structure

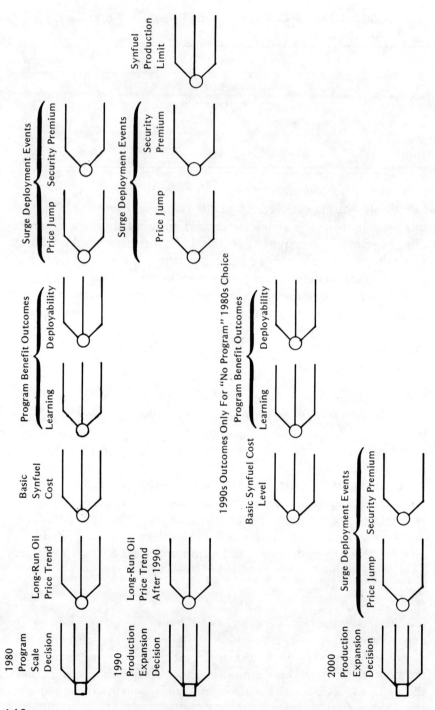

jump events or increases in the long-run security premium could occur acts to increase the expected value of synfuel as compared to oil, thus making larger synfuel production levels more attractive even before the events occur.

Overall Decision Tree

The overall structure of the decision analysis model or tree presents a complicated and imposing picture. Figure 4-3 displayed the structure of decision points at the beginning of each of three decade-long decision epochs. Figure 4-5 displayed the structure of the decision tree for the 1980s decade only. Figure 4-8 extends Figure 4-5 by adding the decision and uncertain outcome nodes for the 1990s and 2000 decades. Figure 4-8 presents the overall structure of the decision analysis tree used for this analysis.

A path through the decision tree must take a single branch or fork at each decision or chance node. Each path defines a particular set of outcomes or a scenario. The structure of the decision tree defines many possible paths. The decision tree in Figure 4-8 systematically defines over half a million different scenarios. Each scenario has a particular likelihood of occurring according to the probability assignments at the various chance nodes.

CALCULATION OF BENEFITS FOR ALTERNATIVE SCENARIOS

Once a particular scenario is fully defined at the end of path through the decision tree, the model calculates net benefits or costs. The calculation of benefits is straightforward. For each year during the life of plants built in a particular decade, the cost of synthetic fuel production is subtracted from the social cost of oil. If synfuel costs exceed the cost of oil, negative benefits or costs accrue. Benefits accrue when oil costs exceed synfuel production costs. Yearly benefits equal this dollar per barrel cost differential multiplied by the volume of production.

Year-by-year benefits or costs are summed with discounting to yield a net present value of production. Since discounting reflects the time value of money, the benefits or costs of earlier years are weighted

Figure 4-9. Illustration of Cost-Quantity Benefit Calculation and Sources of Benefits

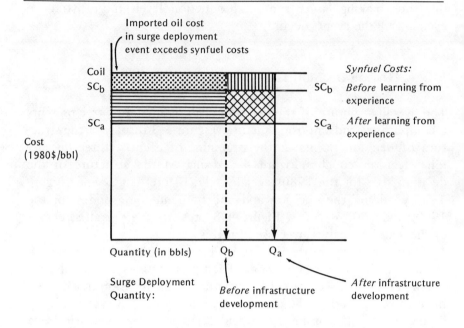

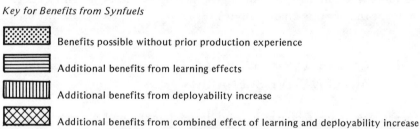

more heavily. Benefits are calculated separately for the group of plants deployed in each decade because learning effects and any factor cost inflation cause synfuel production costs to vary from one decade to the next. Total benefits equal the sum of the net present value (in 1980) of each group of plants deployed in the 1980s, 1990s, and 2000–2010 decades.[18]

Benefit calculations for each scenario reflect three major factors, as illustrated by Figure 4-9. The difference between oil costs and synfuel production costs are the primary source of benefits. Learning effects lower synfuel costs, increasing the difference between synfuel costs and oil costs if the difference is positive, as it should be if a surge deployment is warranted. (The difference is reduced if synfuel costs exceed oil costs.) Deployability effects allow more synfuel plants to be built in each period. This increases the total volume of production over which per barrel savings relative to oil may be obtained. Thus, the benefit calculations reflect not only changing oil prices and synfuel costs but also the values of better plants and more plants central to this assessment of the insurance value of improved strategic capabilities with the synfuels option.

Summary

The structure of the decision analysis model is complex because of the many uncertain parameters and events that significantly affect a systematic assessment of how large a synfuels program is appropriate. The decision analysis model represents a tool that can be used to assess alternative program decisions for an appropriately wide range of parameter and probability values. The model does not provide one answer. Rather, it should be used as a tool to help illuminate the complicated tradeoffs between the size of an initial production effort, contrasting risks of large production subsidies or foregone savings, and varying levels of improved capabilities with the synfuels option.

Next, Chapter 5 illustrates how the model works.

NOTES

1. Benefits and costs for synfuels plants decided upon after the year 2000 and beginning production beyond 2010 are not included in the decision analysis for three reasons. First, synthetic fuels deployment after 2010 occurs far enough in the future to reflect an evolutionary scenario for growth of a synfuels industry rather than the surge deployment concept central to the insurance framework for the benefits of initial synfuels production experience. Secondly, production benefits or costs from facilities built so far in the future are so highly discounted that they do not weigh heavily in net present value assessments. Finally, limiting the problem to three decade-long decision epochs reduces computational scope without sacrificing the essential elements of the decision problem.

2. A large number of branches for each node could become computationally burdensome quickly because the number of paths to be evaluated is the multiplication of the number of branches associated with each decision and chance node proceeding through the decision tree. For a decision tree with two 4-branch nodes, seven 3-branch nodes and four 2-branch nodes, the total number of alternative scenarios evaluated by the decision model is $4^2 \times 3^7 \times 2^4$, or 559,872. Clearly, computational scope increases rapidly as nodes or branches are added to a decision tree.
3. For a more detailed description of the decision analysis model as well as the model's Fortran computer coding, see Appendix IV–A of Harlan (1981).
4. In technical terms, most outcomes are independent of program scale decisions. Other outcomes, particularly the program benefit outcomes, are dependent on previous decisions.
5. The percentage increase approach to modeling this important effect is appropriate since factor cost inflation primarily raises effective capital construction costs for a synfuels plant. Per barrel capital costs are a significant share of total production costs for coal-based synthetic fuels and contribute substantially to the wide range of expectations for synfuel production costs.
6. The calculation assumes a 5 percent discount rate for 1.0 MMBD production coming on-line in 1990. The calculation is:

$$10 \times 1.0 \times 10^6 \times 365 \times (1.05)^{-10} \times 15.4 = 34.5 \times 10^9.$$

The factor 15.4 is the accumulating discount factor for a constant stream over thirty years at a five percent discount rate.

7. Shifts in oil price trajectory from one long-run trend to another are assumed to occur gradually over a period of twenty years as a linear shift between 1990 and 2010.
8. For background on these price trajectories and associated energy supply/demand projections, see Department of Energy (1980b).
9. Some information about basic synfuel production costs might be obtained from a few commercial-scale projects for coal-based synfuels that may proceed even if in the absence of any significant special program to promote initial synfuels production. However, for this analysis we will assume that the basic cost level coal-based synfuel production will remain substantially uncertain in such a "no program" case.
10. For the "no program" case, these program benefit outcomes are not resolved in the 1980s decade but remain uncertain for the 1990s decision. The outcomes are resolved following 1990 decisions to produce some amount of synthetic fuels in the 1990–2000 period.
11. The debate ranges widely from virtually zero learning to over 20 percent. It is important to note that learning with capital-intensive technologies such as synthetic fuels may be less than learning experienced with relatively labor

intensive technologies. Moreover, many of the process operations and equipment in a synthetic fuel plant are commonly applied in other industries. Most learning may occur in innovative sections of a complex plant. For a good discussion of learning with chemical process plants, see Merrow (1978). Merrow's expectations for learning fall at the lower end of many ranges. The optimistic end of the range for learning outcomes is represented by Hirschmann (1964). Hirschmann suggests that a 20 percent learning curve may be applicable to capital-intensive chemical processing technologies also. Few empirical studies of learning in capital intensive chemical process technologies are available. Enos (1962) describes substantial technological improvements in oil refining, but coal synfuels may not necessarily parallel that experience. A recent empirical study of coal-fired power plant construction observed about an 8 percent learning curve. See Ostwald and Reisdorf (1979).

12. For example, a price step event occurring in the 1990s decade (which is modeled to occur in 1999) will increase the value of plants that are decided upon in the year 2000 and coming into production some years (nominally ten years) later. Consequently, more extensive production expansion will be sought for the year 2000 decision. Moreover, the expectation for a price step event in the 1990s will increase the expected lifecycle value of synthetic fuel plants built before the event as part of 1980s or 1990s deployment decisions.

13. For a good review of why the value of the oil import premium might increase with changed probabilities see Hogan (1980) and Department of Energy (1980c). Some developments that might lead to security events are also suggested qualitatively in Deese and Nye (1980).

14. Allowing factor cost inflation to continue for deployment decisions in all decades is a more realistic and economic representation of the supply curve in each decade for increments in synthetic fuels production. If factor cost inflation occurred only in the initial deployment decade, production increment choices in 1990 and 2000 for the decision analysis model would tend to be either maximum or zero. Allowing factor cost inflation in all time period increases makes it more likely that interim levels of production increments for the 1990s and 2000 decisions would be better than either the all (maximum possible) or nothing (zero) choices.

15. For comparisons of Projections, see Table 5-4 in Department of Energy (1981b).

16. For cases where no production is obtained in either the 1980s or 1990s decades, synfuel costs and program benefit outcomes remain uncertain through the year 2000 production expansion decision.

17. Although the analysis does not include the benefits of synfuel plants deployed after 2010, the full lifecycle benefits and costs of the plants built during each of the three decision epochs are calculated. For example, cal-

culations are carried out until the year 2040 for plants that are decided upon in the year 2000, begin production ten years later, and have a nominal life of thirty years.
18. For detailed review of the benefit calculations, see Harlan (1981), especially Appendix IV-A and IV-B.

5 HOW THE MODEL WORKS

Some readers may be unfamiliar with the basic operation of a decision analysis framework and model. Although the model appears to be complex, it primarily performs simple accounting functions. First, the model defines scenarios. Second, it calculates benefits and costs for each scenario using straightforward arithmetic. Finally, it averages benefits or costs across all scenarios with appropriate weights.

CALCULATING BENEFITS AND COSTS FOR A PARTICULAR SCENARIO

At this point, it may be useful to consider a simplified example of how the decision analysis model works. The parametric estimates of benefits and costs developed in Part I of this study may be used to assess the net benefits or costs for a particular well-defined scenario. A particular scenario may be specified by assigning specific values to each of the uncertain parameters required to estimate costs and benefits. With the necessary factors so specified, the appropriate entries from the various parametric tables in Chapters 2 and 3 can be located and combined with other factors to calculate total benefits or costs.

Consider a particular scenario defined as follows:

- Medium long-run oil price trend;[1]
- $50 per barrel basic cost level for synfuels;[2]
- Learning outcomes of 8 percent for technology improvement learning and 5 percent for selection learning;
- A price jump event in 1993;[3]
- A surge deployment effort of forty nominal plants (2.0 MMBD) initiated in 1993 with plants beginning production in 2000;
- Ten of the forty plants built made possible because of enhanced deployability realized via initial production experience; and
- Following the surge deployment in the 1990s, smooth long-run oil prices and no further surge deployment.

Referring to Table 2-1, selection learning for a synfuel cost outcome of $50 per barrel yields net present value (NPV) benefits of $0.3 billion per plant or, for forty plants, $12 billion. Referring to Table 2-2, cost savings from technology improvement learning amount to about $2.50 per barrel. Using Table 2-3, we can calculate that this $2.50 per barrel saving accumulates over forty surge deployment effort plants to about $9.6 billion NPV ($2.50 × 0.096 × 40 = 9.6).

Benefits from enchanced deployment capability are estimated by reference to Table 2-6. For a 1993 price jump above the medium long-run trajectory, each surge deployment plant yields life cycle net present value benefits of about $0.7 billion. (This per plant benefit estimate is the interpolated result for $47.50, reflecting a $50 synfuel cost outcome less $2.50 for learning curve cost reductions.) Since enhanced deployability resulted in ten additional surge deployment plants being possible, net incremental benefits from enhanced deployability are about $7 billion. Total benefits are the sum of benefits from learning (12 + 9.6 = 21.6) and enhanced deployment capability (7.0), or about $29 billion net present value (21.6 + 7 = 29).

To continue the illustration, assume the improved capabilities that yield these surge deployment benefits were obtained via an initial production program comprised of ten plants initiated in the 1980s and beginning operation in 1990. The expected life cycle subsidy costs for the ten initial production plants is $9 billion, NPV.[4] (See Table 3-9 value for $50/bbl and medium oil price trend. No factor cost inflation is assumed for this example.) Thus, for this particular scenario, the benefits ($29 billion) of an initial synthetic fuel production program exceed costs ($9 billion) by a substantial margin. Net benefits for this particular scenario are about $20 billion.

However, for other scenarios, costs could exceed benefits. Consider, for example, a scenario characterized by the $60 per barrel outcome for synfuel costs with the low oil price trajectory and no surge deployment event. Under such a scenario, each initial production plant starting up in 1990 would incur (based on Table 3-9) net present value costs for production subsidies of about $4.4 billion. A ten plant program would incur a cost of $44 billion. Since no price jump event occurs, however, the offsetting benefits from learning or deployability for surge deployment efforts would be negligible.

The decision analysis model used in this analysis performs the function of an accounting mechanism. First, the model systematically defines each alternative scenario implied by the range of possible combinations of outcomes for uncertain parameters. Net program benefits or costs are then calculated for the specific scenarios or paths through the decision tree. Conceptually, this process is analogous to defining the set of parameters that define a specific case to be assessed, locating particular values in the parametric benefit and cost tables presented in Chapters 2 and 3, and then combining those table values as appropriate to calculate net benefits or costs. This process is repeated for each possible scenario.

WEIGHTING SCENARIO ESTIMATES BY PROBABILITIES

If these two example scenarios—net benefits of $20 billion or net costs of $44 billion—were the only two possible scenarios and each occurred with 50 percent probability, the net expected benefits of an initial production program of ten plants would be a negative benefit, or cost, of about $12 billion (50%(20) + 50%(-44) = -12). However, the expected value of an initial production program is related to an enormous number of possible scenarios. Each scenario is defined by a particular combination of outcomes for uncertain parameters and related decisions regarding future synfuel production.

"Expected" benefits must reflect outcomes for all possible scenarios. In simplest terms, the decision analysis model determines net expected benefits by weighting the outcome for each scenario by its probability of occurring and adding these probability-weighted outcomes together for all scenarios. This process allows extreme scenario outcomes, which might occur with relatively low probability but have large absolute benefits or costs, to be integrated appropriately into expectations for program benefits and costs.

For example, if the $44 billion cost of our previous example was a scenario occurring with only 10 percent probability, while the $20 billion benefit scenario occurred with 90 percent probability, net expected program benefits, although still positive, would be reduced from $20 to $14 billion (10%(-44) + 90%(20) = 13.6). This capacity of decision analysis to explicitly and systematically integrate a range of uncertain and potentially extreme scenario outcomes makes decision analysis particularly appropriate for the assessment of the insurance benefits of initial synfuel production efforts.

COMPARING APPROPRIATE PROGRAM SIZE FOR TWO EXTREME SCENARIOS

The operation of the decision analysis model for addressing the "How large?" question may be illustrated by considering two extreme scenarios—one favorable to a large synfuels program and another relatively unfavorable.

The appropriate size for an initial synfuel production effort depends on expectations for long-run oil price trends and synfuel production costs. In a scenario favorable for synfuels, where oil costs are expected to exceed synfuel production costs, the benefits of larger programs should increase in direct proportion to production volume. Moreover, infrastructure development effects might cause larger programs to yield even greater benefits, since larger production volumes could be attained. (Cost savings from learning would accrue over a larger volume of production.)

Other effects act to offset the larger benefits of larger programs, even in scenarios favorable to synfuels. Factor cost inflation for larger initial production efforts might more than offset the added production and deployability benefits. Synfuel production limits might make the additional infrastructure development from larger initial production programs unnecessary.

Figure 5-1(a). Results for Two Contrasing Scenarios

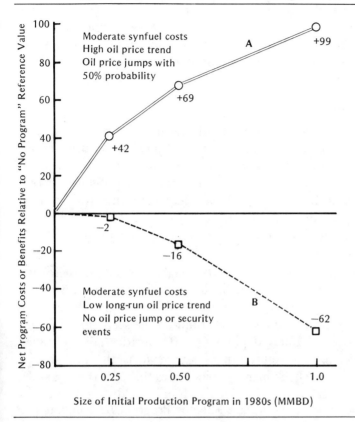

Moreover, scenarios where synfuels yield large benefits may be offset by scenarios (such as high synfuels cost, low oil price, or no surge deployment events) where large initial production efforts incur large subsidy costs. The value of additional deployment capability purchased via larger initial production efforts would be small, since capacity expansion in 1990 or 2000 might be small or zero. The increasing benefits to larger programs in scenarios favorable to synfuels should be weighed against the risks of significant losses in scenarios unfavorable to aggressive synthetic fuel deployment.

Figure 5-1 illustrates example results from the decision analysis model for two particular combinations of outcomes for long-run oil price trends, the basic synfuel cost level and the probability of events. One scenario is favorable to aggressive synfuel deployment; the other is unfavorable.

Curve A in Figure 5-1(a) denotes a scenario (A) of moderate synfuel costs, a high long-run oil price trend, and 50 percent probability for price jump events. Net benefits or costs are measured relative to a reference value of no initial production in the 1980s and graphed against program size. For Scenario A, the net benefits of small (II), moderate (III), and large (IV) programs are +42, +69, and +99 respectively.[5]

Curve B in Figure 5-1(a) depicts net benefits for a combination of outcomes unfavorable to synfuels deployment—moderate synfuel costs, low long-run oil prices, and no price jump or security events (Scenario B). Programs II, III, IV each have negative benefits or net costs of -2, -16, and -62 respectively. The "no program" choice would be technically the best if we were sure that this scenario would be the only possible one. In such a scenario, negative benefits or costs for production subsidies increase on the margin with increasing program size.

Figure 5-1(b) shows the probability-weighted result if these two scenarios each occur with equal likelihood. The negative quantities associated with the scenario (B) unfavorable to synfuels are subtracted from the positive benefits accruing in the scenario (A) favorable to aggressive synfuels deployment. (Quantities are multiplied by probabilities before subtracting.) Curve R is the probability-weighted result. The check mark ($\sqrt{}$) indicates that, for this example, program III is clearly the "optimal" choice yielding the greatest expected value. (Net benefits are +26 for Program III compared to +20 for Program II, or +18 for Program IV.)

Figure 5-1(c) displays the net result if Scenario A is twice as likely to occur as Scenario B. With those relative likelihoods for these two contrasting scenarios, negative benefits for Scenario B are reduced by one-half since it is believed to be only one-half as likely to occur as the Scenario A. Curve S is the probability-weighted result. The program choice yielding the greatest expected value has shifted from Program III (with expected net benefits of +41) to Program IV (with expected benefits of +45).

RELATING THESE EXAMPLES TO THE "FULL" MODEL RESULTS

This chapter has illustrated how the decision analysis model works for two contrasting scenarios. The full model systematically defines,

Figure 5-1(b). Probability-Weighted Result for Two Extreme Scenarios Each Having Equal (50%) Probability

Figure 5-1(c). Probability-Weighted Result When High Synfuel Benefits Scenario Is Twice As Likely to Occur.

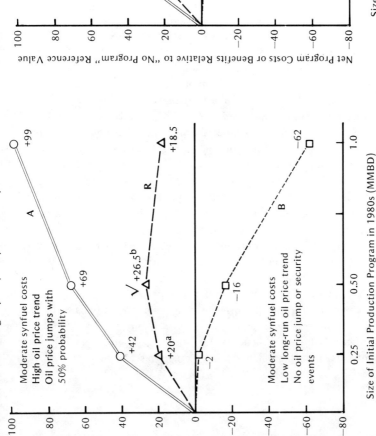

[a]Calculation for Program II: 50% (+42) + 50% (−2) = 21 + (−1) = +20
[b]Calculation for Program III: 50% (69) + 50% (−16) = 34.5 + (−8) = +26.5
[c]Calculation for Program IV: 50% (99) + 50% (−62) = 49.5 + (−31) = +18.5
[d]Calculation for Program II: 2/3 (42) + 1/3 (−2) = 28 − 2/3 = 27 1/3
[e]Calculation for Program III: 2/3 (+69) + 1/3 (−16) = 46 − 5 1/3 = 40 2/3
[f]Calculation for Program IV: 2/3 (+99) + 1/3 (−62) = 66 − 21 = 45

evaluates, and intergrates many contrasting scenarios. The model calculates net benefits or costs for each scenario and combines estimates for both average and extreme scenarios with weights that reflect the decisionmaker's probabilities for various events and parameter values.

The results in the next chapter are based on comparisons and calculations analogous to those outlined above. Model results reflect the probability-weighted outcome ("expected value") from the systematic comparison of over a half million different scenarios. Results are the expected net benefits or costs of each alternative program choice for the 1980s decade.

NOTES

1. Department of Energy, Policy and Fiscal Guidance Planning Cases, as described in Chapters 2 and 3. Other terms are also defined in Chapters 2 and 3.
2. Form value should be reflected in this synfuel cost outcome. For simplicity, this example will assume no oil import premium, since this would complicate illustration of the method for estimating the value of learning cost savings.
3. A price jump event is modeled as a doubling of oil prices followed by a linear decline over fifteen years back to an underlying long-run oil price trend.
4. In fact costs for initial production plant subsidies under this scenario would be less than $9 billion. The price increases that motivated the surge deployment would reduce the lifecycle subsidy costs for the initial program plants initiated in the 1980s and beginning production in 1990. For simplicity, in order to utilize the tables in Chapter 3 for the explication of this example, we suppress this effect. Under the price jump scenario, the subsidy cost for initial production plants beginning in 1990 at $50 per barrel is $-1.4 billion NPV (no subsidy), rather than the estimate of $0.9 billion NPV per plant utilized above. These complications illustrate the value of using the decision analysis model for making cost-benefit comparisons rather than attempting to extract values from the corresponding parametric tabulations.
5. Note the decreasing slope of Curve A in Figure 5-1(a). Although there are significant additional benefits for the largest program size, net benefits exhibit declining marginal returns to program scale.

6 THE DECISION ANALYSIS RESULTS

USING THE DECISION ANALYSIS TOOL

A decision analysis approach allows many uncertain and contrasting decision factors to be addressed systematically in the assessment of "How large?" an initial synthetic fuel production effort is appropriate. It is not suggested, however, that the output of the decision analysis model should drive any particular decisions. The model results merely illuminate a correct analytical decision. The correct public policy decision should consider the analysis results but must reflect other considerations as well. For example, budget constraints and competing priorities may keep us from buying as much strategic insurance as we might think appropriate. The decision analysis approach and model are useful as tools that can assist decisionmakers with some of the complex choices they face.

The decision analysis results presented in this chapter provide signposts for decisionmaking. The results show that our decisions on how much insurance to purchase should be guided by our expectations for key uncertainties, including future oil price trends, synfuel production costs, and surge deployment events. The decision analysis model can be used as a tool to test the implications of alternative assumptions or "world views" regarding those key uncertainties. Also, rather than focusing on the appropriate size of an initial production program, a decision analysis model and approach can help assess the advantages of emphasizing learning benefits or infrastructure development for initial production programs of similar size. Alternative program strategies or portfolios can be assessed. Finally, a decision analysis approach can help characterize the relative gains or losses of a phased or sequential approach as compared to a strategy of irrevokable commitments now, in spite of remaining uncertainties.

The decision analysis results presented in this chapter should clarify issues concerning the appropriate levels of social investment in initial commercial-scale production of coal-based synthetic fuels. Such large investments of the nation's limited resources should be prudently considered. Another objective of this study is to illustrate the use of a decision analysis approach as a tool that is applicable to similar decision problems, both within the energy area and elsewhere.

Chapter 6 is divided into two major sections. The first section describes and interprets the results for the base case parameter and probability specifications outlined in Chapter 4. The second section presents results for six groups of sensitivity cases. These cases generally vary one parameter or probability assumption at a time from the values assumed for the base case. Chapter 7 concludes Part II with summary observations on the "How large?" question drawn from the decision analysis results presented in this chapter.

BASE CASE RESULTS

Questions Addressed by Model Outputs

The basic output of the decision analysis model is a numerical calculation of the expected net benefits of alternative initial synthetic fuel production programs. Benefits or costs are discounted using a 5 percent real discount rate to calculate a net present value in 1980. All calculations utilize constant real 1980 dollars.[1] A technical interpretation of the results is that the "best" or "optimal" program choice is the one with the highest expected net present value.

The program choice recommended as technically the "best" is only one item of information from the model results. For a given sensitivity case, the results also illuminate how much better one program is than another. Moreover, the results of the decision analysis model can be used to illustrate how the appropriate program

Table 6-1. Expected Benefits of Initial Synfuel Production Programs for Base Case (net present value in billions of 1980$ at 5 percent discount rate).

	Program I	Program II	Program III	Program IV
Amount of production in 1980s	0.0 MMBD	0.25 MMBD	0.5 MMBD	1.0 MMBD
Qualitative label	No Program	Small	Moderate	Large
Total expected value	37.8	69.4	81.9	83.0
Net expected value relative to "no program" reference value	(37.8 = 0)	+31.6	+44.1	+45.2

choices may change with different expectations for key uncertainties, such as synfuel costs or oil prices. The extent to which they do change provides an indication of the value of obtaining information about those key uncertainties. For example, the model results aid in assessing such questions as: "If I am certain that synfuel costs are high (or low), what sort of initial production program looks best?"

This section describes results and investigates such questions for the base case parameter and probability values specified in Chapter 4. Base case results are discussed in some detail in order to illustrate how the model results can be used and interpreted.

Basic Results

Table 6-1 displays the output of the decision analysis model for base case parameter and probability values. Benefits are expressed as billions of dollars (1980$) net present value. For this base case run, the moderate size initial production program (Program III or 500,000 barrels per day in the 1980s) and the large program (Program IV or 1.0 million barrels per day in the 1980s) yield essentially the same level of net benefits. A program twice as large yields less than a 2 percent increase in expected net benefits.

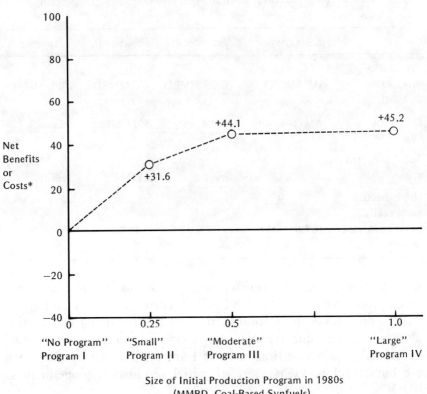

Figure 6-1. "Standard Base Case" Results

*Net program Benefits or costs relative to "No Program" reference value (Net Present Value in billions of 1980$ at 5% discount rate.)

The last line of Table 6-1 shows *net* program benefits expressed with the "no program" outcome establishing as a zero value reference point. The total benefits associated with the "no program" choice in 1980 are subtracted from the total benefit estimates of other programs to determine net benefits. A decision to forego production in the 1980s (the "no program" choice) does not preclude the option for significant synfuels deployment in either the 1990s and/or the 2000-2010 decades. That choice merely defers by at least a decade the benefits of initial practical production experience.

Figure 6-1 is a graphical display of the results of the decision analysis for the base case. Initial production level is on the horizontal axis. Net program benefits are on the vertical axis. By definition, the "no program" choice yield zero net benefits. This display illustrates the substantial benefits from a small initial production program (0.25 MMBD) in the 1980s, a significant increment (40 percent) in benefits for the moderate size (0.5 MMBD) program, but only slightly increased net benefits for the large initial production program (1.0 MMBD by 1990) relative to the moderate size effort. Net program benefits clearly exhibit declining marginal returns to increasing program size.

A Qualitative Interpretation of Results

These declining marginal returns to increasing program size are due in part to the risks of substantial production subsidies across a large volume of production if synfuel production costs are high and long-run oil prices are low or no events occur. (Recall Scenario B in Figure 5-1.)

The substantial benefits of a small initial production program reveal the value of information about uncertain synfuel production costs. Initial production experience will reduce some of the uncertainty about which scenario is more likely to occur. Subsequent deployment expansion decisions can be made on an improved information basis. Deployment expansion decisions in 1990 can be aggressive if synfuel costs are likely to be lower than oil costs. Conversely, if synfuel costs prove relatively high, deployment expansion can be appropriately small or zero. The ability to make better deployment decisions in 1990—exercising opportunities where benefits are likely or going slowly where we are less sure or where costs are likely—is the basic source of information benefits from initial production experience.

The substantial value of information illustrates the advantages of a phased approach to deployment decisions. The uncertainty about basic synfuel production costs will be reduced with an initial production program of any size. However, large initial production efforts involve major commitments in the face of substantial cost uncertainty. Losses may be great if synfuel costs turn out to be high. Smaller initial production efforts yield essentially the same information about synfuel production costs but incur lower expected subsidy costs. An initial production strategy emphasizing the value of improved information on synfuel production costs would reflect a phased or sequential approach. A small initial production effort oriented toward producing cost information could be followed by production expansion decisions reflecting improved information about key uncertain parameters.

Other considerations complicate the design of an initial production program. The relationship of other sources of benefits to program size should also be considered. For example, larger initial production efforts develop the infrastructure more extensively expanding the range of future choices. Greater surge expansion of synfuels production capacity should be possible if information about production costs proves favorable to aggressive deployment.

On balance, the base case decision analysis results suggest an intermediate strategy. Expected net benefits from the small (250,000 B/D) program are significant relative to a "no program" reference case. These benefits reflect the value of information and initial installment of learning. Benefits increase by about 40 percent when the program size is doubled from 0.25 MMBD to 0.5 MMBD. The added learning and deployability benefits make increasing program size worthwhile, despite the fact that the moderate program experiences 10 percent factor cost inflation. Expansion of the initial program to 1.0 MMBD yields few additional expected benefits. The risks of large production subsidies and the added near-term costs from substantial (25 percent) factor cost inflation offset the additional deployability, learning, and production benefits. Figure 6-2 summarizes this qualitative interpretation of the shape of the net program benefits curve for the base case results.

Results for Particular Synfuel Cost and Oil Price Outcomes

The base case results reflect different assumptions about the probabilities for various synfuel cost and oil price outcomes. One obvious question concerns how relative program choices might change for scenarios where particular oil price or synfuel cost outcomes were certain to occur. Table 6-2 shows relative program values for cases where synfuel production costs are known with certainty but oil prices are uncertain. Figure 6-3 presents some of these results in graphical form. Table 6-3 presents similar results for cases where

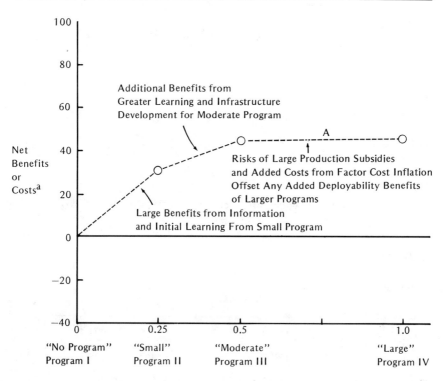

Figure 6-2. Qualitative Interpretation of "Standard Base Case" Results

[a] Net Program Benefits Relative to "No Program" Reference Value (NPV in Billions of 1980$ at 5% Discount Rate)

particular long-run oil prices occur with 100 percent probability but synfuel costs are uncertain, reflecting the base case probability distribution. (Note that although Table 6-3 values reflect certainty for a particular long-run trend, price jump events still occur with 50 percent probability in each decade.) Such results can be used to assess relative program values for probabilities of oil prices and synfuel costs different than the particular probabilities assumed for the base case. For example, if one has a 0-50-50 percent expectation for the low-medium-high ($35, $58, $65 per barrel) synfuel cost outcomes (rather than the base case probabilities of 20-45-35), relative program

Table 6-2. Relative Value of Alternative Initial Production Programs for Particular Synfuel Cost Outcomes[a]

Certain Synfuel Cost Level	I (= reference value)	II	III	IV
(1980$/bbl)				
Low ($35/bbl)	(111.1 = 0)	+42.9	+81.6	+139
Mid ($50/bbl)	(41.7 = 0)	+29.8	+45.0	+51.6
High ($65/bbl)	(8.1 = 0)	+10.2	+4.1	−33.6

[a] Boxes denote the best program choice.

Table 6-3. Relative Value of Alternative Initial Production Programs for Particular Long-Run Oil Price Outcomes.[a]

Certain Long-Run Oil Price Trajectory	I (= reference value)	II	III	IV
Low	(19.3 = 0)	+20.5	+21.1	−0.2
Medium	(37.2 = 0)	+31.2	+43.6	+44.6
High	(57.7 = 0)	+43.2	+67.8	+91.9

[a] Boxes denote the best program choice.

values can be calculated by simply averaging results for the medium and high synfuel cost outcomes shown in Table 6-2.[2]

How Much Is Better Information Worth?

Tables 6-2 and 6-3 and Figure 6-3 illustrate how appropriate program choices and the level of expected benefits change with different outcomes for key uncertainties, such as the synfuel costs or the long-run oil price trend. Advance information that clarifies the best decision can be valuable to decisionmakers, because they can pick the program choice which is best given sure information about

Figure 6-3. Base Case Results for Particular Oil Price Trends and Synfuel Cost Outcomes

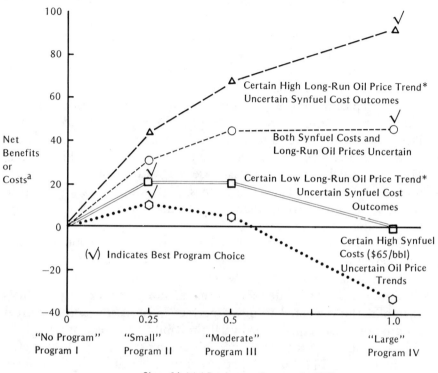

ªNet Program Benefits Relative to "No Program" Reference Value (NPV in Billions of 1980$ at 5% Discount Rate)

*High or low refers to long-run oil price trend only; oil price jump events occur with 50% likelihood in each decade.

---

critical decision factors. Losses from poorer choices that might have been made without improved information can be avoided.

The base case decision analysis results in Tables 6-2 and 6-3 illustrate the potentially substantial value of information about synfuel production costs. For certain "high" synfuel costs, Program II is the best choice. If synfuel costs are uncertain, Program IV is the best choice. Moreover, for high synfuel costs, the small program (II) is better than the large program (IV) by approximately $44 billion net present value (NPV).[3] Advance information about high synfuel production costs should change the decision on program size and avoid losses that would make the decisionmaker $44 billion worse off.

The above example suggests that better information about synfuel production costs prior to a decision on 1980 program size could be quite valuable. However, information that some outcomes were certain to occur would not necessarily change decisions. In our base case example, if we learn for sure that synfuel production costs are "low" or "medium," then the optimal program choice remains Program IV. The decision after the information is obtained is the same as the choice that would have been made before the information was available, given our base case specifications for the probabilities of various synfuel cost outcomes. If decisions do not change in response to information, then the information, at least technically, has no value.

The expected value of perfect information about some uncertain parameter depends on how likely a particular uncertain outcome is and how much better off we become by learning about that particular outcome before a decision is made. For our base case synfuels example, the expected value of perfect information about synfuel production costs equals our initial probability for realizing the high cost outcome (35 percent) multiplied times the amount benefits are increased ($44 billion) because the decision changes (from a large to small program) in response to the information that synfuel costs are high for certain. If perfect information about the basic synfuel cost level were available before the 1980s production decision, it would be have an expected value of about $15 billion NPV (35% × $44 billion = $15 billion).

In practice, perfect information about basic synfuel production costs is unlikely to be available before decisions on initial production programs must be made. Good information about synfuel costs may require practical production experience. Moreover, rarely is information perfect in the sense of providing 100 percent certainty about an outcome.

Nevertheless, an estimate of the value of perfect information about synfuel production costs is useful for several reasons. First, the calculation illustrates the value of information for later phased decisions about synfuel deployment levels. Initial production efforts should provide improved information for 1990 and 2000 deployment expansions decisions. Although the example calculation illustrates the expected value of perfect information for the 1980 synfuels production decision, analogous calculations are possible for the 1990 and 2000 deployment decisions. Those phased production commitments can respond to information about major uncertainties, avoiding costs or exercising opportunities.

Secondly, the expected value of perfect information provides an estimate of the value of a flexible approach to production goals that responds to developing information about uncertain but critical decision factors. Technically, the base case results suggest that we should proceed with a large initial production program. However, that recommendation is predicated on the base case probability specifications for synfuel costs. If, as initial production projects are developed and become better defined, synfuel production costs appear increasingly likely to be high, the appropriate program size shifts toward smaller initial production efforts. If we believe that developing information about synfuel costs could show that they are certain to be "high" ($65 per barrel or more), the expected value of information illustrates the substantial gains from allowing production objectives to change in response to improved information. Stated differently, the expected value of perfect information illustrates the maximum expected cost of a production strategy and commitment that ignores developing cost information and sticks inflexibly to production levels chosen prior to the availability of the new cost information.[4]

Thirdly, the expected value of perfect information benchmarks the expected losses of proceeding with production commitments without first obtaining improved (in fact, perfect) information about synfuel production costs. For example, suppose we could defer decisions on initial production goals until after a single initial project had clarified the basic synfuel cost level. Then the expected value of information quantifies the maximum value of the information from the initial project. It also indicates the expected costs of proceeding with production decisions and commitments without waiting for the cost information from the initial project.

The expected value of perfect information on synfuel costs is a substantial, but not overwhelming, number. Its magnitude may be put into perspective by noting that $15 billion net present value is about one-third to one-half of the total benefits estimated for initial production efforts.[5]

Table 6-4. Relative Program Benefits for Various Combinations of Long-Run Oil Price and Synfuel Cost Outcomes.

Synfuel Cost Outcome	Oil Price Outcome	Joint Probability of Combination in Base Case Specifications	Program Level			
			I	II	III	IV
Low	Low	5.0%	81.1	114	140	☐172
Low	Mid	10.0%	111	154	193	☐250
Low	High	5.0%	141	194	246	☐328
Mid	Low	11.2%	21.5	39.1	☐42.3	25.9
Mid	Mid	23.5%	40.5	70.3	85.6	☐91.9
Mid	High	11.2%	64.3	106	134	☐164
High	Low	8.8%	☐1.6	−1.6	−18.8	−77
High	Mid	17.5%	6.9	☐16.9	10.6	−27.3
High	High	8.8%	16.8	41.0	☐46.4	29.7

[a] Boxes denote best program choice.

Preferences for Particular Views of Oil Prices and Synfuel Costs

Table 6-4 displays expected values for specific combinations of long-run oil price trajectories and basic synfuel cost outcomes. Results in boxes are the program choices yielding the largest benefits. The third column in the table denotes the joint probability for the specified combination of synfuel production costs and long-run oil price trajectories (assuming base case probability specifications).

Although this table may appear complicated, several general observations may be readily drawn. Preferred program decisions change with different oil price and synfuel cost outcomes. Of nine possible combinations, Program IV is preferred for five combinations, Program III for two combinations, and Programs I and II for one combination each. If synfuel costs are believed to be low ($35/bbl) or moderate ($50/bbl), Program IV is recommended, unless the underlying long-run oil price trajectory is also low. If synfuel costs are expected to be high ($65/bbl or more), Program IV is never recommended. Programs I, II, and III are preferred for low, medium, and high long-run oil price trajectories, respectively.

Table 6-4 illustrates some possible reasons for the wide divergence of opinion regarding the expected costs and value of initial synthetic fuel production efforts. Technological optimists, who expect synthetic fuel costs to fall in the low to middle range, would favor Program IV or even larger initial production efforts. In contrast, the preferred program choice of those expecting synfuel costs to be high ($65/bbl) varies directly with long-run oil price expectations. (Note that Table 6-4 results include a 50 percent probability for a price jump event above the long-run trend in each decade.) Clearly, many individuals have varying expectations (often only implicitly rather than explicitly defined) for these two uncertain parameters, which may be dominant in shaping individual opinions of the merits of initial synthetic fuels production efforts.

RESULTS OF SENSITIVITY ANALYSIS

The Key Role of Sensitivity Analysis

One major advantage of decision analysis is that uncertainties about major parameters and events can be explicitly and systematically reflected in the analysis. The base case results derive from probability and parameter values reflecting one view of the world regarding uncertain outcomes and their probabilities. Strictly speaking, of the four alternative levels of initial production activity examined, the decision analysis recommends Program IV—the largest level examined—as the best or optimal choice. However, Program III—representing an initial production level half of that for Program IV—yields net expected benefits only 2 percent less than the larger program. Also, Program II—a 250,000 barrels per day effort in the 1980s—yields net expected benefits 70 percent as large as expected benefits of the four-fold larger effort represented by Program IV.

Given a 2 percent difference, many decisionmakers would be hard pressed to choose between program levels III and IV. For the base case assumptions, there appear to be no substantial advantages or disadvantages associated with picking one program over the other. However, alternative assumptions for the value of some probabilities or parameters might shift relative program preferences. For example, a different probability for price jump events might shift preferences between Programs III and IV. Lower probabilities for surge deployment events would favor smaller programs relative to larger programs.

In general, alternative assumptions for probabilities and parameters for a variety of factors—such as oil price trajectories, synfuel costs, price step events, and factor cost inflation—change relative program values in intuitively expected directions. For example, if factor cost inflation for Program IV is likely to be greater than the 25 percent assumed in the base case, intuitive reasoning suggests that Program IV becomes less attractive than Program III.

Estimation of the magnitude of changes in relative program benefits, however, is substantially less intuitive and may require separate sensitivity experiments with the decision analysis tool. For example, it is difficult to determine intuitively the probability of a surge deployment event for which a small program becomes better than a moderate program. Sensitivity experiments enable examination of the magnitude of changes in relative program benefits as assumptions change for particular probabilities and parameters. Parameters may be varied from base case assumptions individually or in combinations. Such sensitivity experiments help illustrate how robust various recommended program choices are to changing assumptions. The best program choice may be more sensitive to variations in some parameter and probability assumptions than to others. Assumptions having a strong influence on the appropriate program size help define priority subjects for further analysis.

Overview of Groups of Sensitivity Experiments

Sensitivity experiments illuminate the competing effects that underlie the base case results. They can also illustrate how answers to the "How large?" question change with different probability, parameter, or modeling assumptions. In general, each sensitivity case varies only one parameter or one set or probabilities at a time from those specified for the base case. The sensitivity experiments fall into six groups.

The first group of sensitivity experiments examines changing assumptions for the likelihood of an oil price jump (SDE-P) or security-motivated (SDE-S) surge deployment event. Expectations for the likelihood of surge deployment events are highly subjective. Given the insurance character of surge deployment benefits, the appropriate amount of investment in initial production experience is shown to be quite sensitive to probabilities for surge deployment events.

A second group of sensitivity experiments examines the relative importance of various sources of program benefits. Different sources of benefits have different implications for program size: information benefits may be obtained with a very small program; a moderate size program may yield the majority of learning benefits. Where deployability effects are important, larger initial production efforts may be warranted in order to build substantially more deployment capability. Parameter and probability assumptions in the decision analysis model can be adjusted to generate sensitivity cases that illuminate the interaction of these major sources of benefits.

Other groups of experiments examine the sensitivity of model results and relative program values to different parameter assumptions or modeling relationships. Although decision analysis allows many key parameters to be explicitly and systematically treated as uncertain outcomes, fixed values must be assumed for some parameters or relationships. Experiments with the decision analysis model can illustrate the sensitivity of results to parameters and relationships that are not modeled as uncertainties.

In particular, a third group of sensitivity experiments considers alternative descriptions for factor cost inflation. Factor cost inflation has a strong effect on relative program benefits and costs. Although the amount of factor cost inflation for various program sizes is difficult to estimate, the level of factor cost inflation is not explicitly modeled as an uncertainty. Specific values are assumed for various program sizes. Thus, sensitivity experiments to illustrate changes in relative program benefits for varying amounts of factor cost inflation are particularly important. Results of sensitivity experiments can be used to characterize the expectations for various levels of factor cost inflation which could change the preference ordering of various program choices.

The decision analysis model necessarily assumes a particular stylized description for a surge deployment event. Benefits and costs vary as the description of events changes. A fourth group of sensitivity cases examines alternative descriptions for oil price jump and security events.

The fifth group of cases reviews assumptions that affect the value of the added infrastructure development realized via larger programs. For example, the quantitative linkage between production experience and expanding deployment capability—a highly speculative parameter assumption—is an important determinant of deployability benefits. Deployability benefits—and, hence, the special advantages of larger programs—are also sensitive to assumptions regarding the maximum mid-term market role for coal-based synfuels, the synfuel production limit.

Finally, a sixth group of sensitivity experiments addresses the issue of the possible divergence between (1) the calculation of benefits and appropriate decisions for society as a whole and (2) the choices private firms might make responding to the profits they could realize from initial synfuels production projects. The appropriate level of initial production experience for society as a whole could be different than the levels of initial production investment one might expect from private firms acting in response to their own profit objectives and market opportunities. If such a divergence occurs, government action to encourage additional investment in initial production projects may be warranted.

The discussion of these various sensitivity cases follows a parallel format. A review of background and purpose for the sensitivity experiments is followed by tabular presentation of numerical results for relevant cases. These numerical results display expected net benefits for three different initial production levels measured relative to the reference value of no production in the 1980s. Figures are also used to display results graphically for many sensitivity comparisons.

Results of sensitivity cases are not reviewed at the same level of detail as the results for the base case.[6] However, the detailed information developed for base case results (such as relative program values for particular oil price and synfuel cost outcomes) can also be developed for sensitivity case results.

Following the review of sensitivity case results, Chapter 7 concludes Part II by summarizing some general observations about the decision analysis results. The results of the sensitivity cases help define which probabilities, parameters, and relationships have the strongest influence on appropriate program choices. Results of the sensitivity cases also may help identify a program choice that is relatively "robust"—that is, a choice that, if not the best choice, is nearly as good as the best choice—across many different values for key probabilities and parameters.

Varying Probabilities for Surge Deployment Events

As for any insurance problem, expectations for the events that might motivate surge deployment have a strong influence on the amount of insurance that should be purchased via initial production efforts. Probabilities for surge deployment events are based on highly subjective assessments of the likelihood of developments that may lead to either a stylized price jump event (*SDE-P*) or a substantial increase in the security component of the oil import premium (*SDE-S*).

It is important to recall that both the price jump and security premium events must be relatively long-lasting developments if synthetic fuels production capacity is to be usefully deployed in response to the events. Synthetic fuels are a long lead-time and long lifecycle option for which deployment decisions should be assessed relative to long-term trends or relatively long-lasting events. A temporary price spike less than several years in duration, or a brief, but temporary, disagreement with a major oil producing country are unlikely to be developments that should motivate surge deployment.

Individual expectations for such long-term developments in the face of a complex and uncertain world vary considerably. It may seem difficult or artificial to define probability for the stylized events modeled in the decision analysis assessment. Nevertheless, we should not avoid characterizing our expectations for important developments in world oil markets simply because of the inherent uncertainties and difficulties of the task. Our decisions now should be guided by our best judgments about the likelihood of various future developments, both favorable and unfavorable.

Changing Probabilities for Security Events. Sensitivity cases covering a range of different probabilities for surge deployment events provide an indication of how preferred program choices change with

Table 6-5. Sensitivity Cases Varying Probability for Security-Motivated Surge Deployment Events (*SDE-S*) (net expected program benefits relative to "no program" reference case; billions of 1980$ net present value at 5 percent discount rate).[a]

Sensitivity Case	Probability of SDE-S	Program Level			
		I	II	III	IV
		("no program" reference value)			
A. Base case (SDE-P occurs with 50% probability)	10%	(37.8 = 0)	+31.6	⎡+44.1⎤	+45.2
B. No events; smooth future (no SDE-P or SDE-S events)	0%	(7.5 = 0)	⎡+10.6⎤	+2.5	−34.6
C. No price jump events (SDE-P)	10%	(13.5 = 0)	⎡+14.1⎤	+9.2	−22.2
D. No price jump events (SDE-P)	30%	(23.0 = 0)	⎡+20.2⎤	+20.6	−1.4

[a]Boxes denote best program choice.

one's world view. Table 6-5 summarizes decision analysis results for a set of model runs in which the only parameters varied from base case specifications are the probabilities for surge deployment events. Case *A* is the standard base case, which reflects a 50 percent probability of a price jump (*SDE-P*) event in each decade and a 10 percent probability for a security event (*SDE-S*).

Case *B* reflects a smooth, no-event world where neither price jump events nor security events occur. Both *SDE-P* and *SDE-S* events have zero probability. For such smooth, gradual change, the preferred level of initial production is Program II. Moreover, Program IV has substantial losses ($35 billion NPV) relative to the reference case of no initial production in the 1980s.

In sensitivity Case *C*, the smooth world of Case *B* is complicated slightly by introducing a 10 percent probability for a security event

Figure 6-4. Sensitivity Cases Varying Likelihood of Security-Motivated Surge Deployment Event (SDE-S)

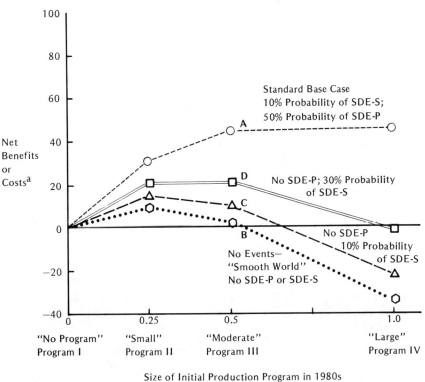

[a] Net Program Benefits Relative to "No Program" Reference Value (NPV in Billions of 1980$ at 5% Discount Rate)

(SDE-S). This slight chance for a security event increases the expected benefits of all programs. The best program choice, however, remains Program II.

Case D increases the probability of a security event from 10 percent to 30 percent. (Still, no price jump events occur.) The best choice shifts from an advantage for Program II (in Case C) to a situation where Programs II and III yield approximately the same level of expected benefits (Program II = +20.2; Program III = +20.6). Figure 6-4 presents results for these sensitivity cases graphically with the program benefit curves labeled A, B, C, or D corresponding to the various case definitions.

Table 6-6. Sensitivity Cases Varying Probability for Price Jump Events (net expected program benefits relative to "no program" reference case; billions of 1980$ net present value at 5 percent discount rate.[a]

Sensitivity Case	Probability of SDE-P	Program Level			
		I	II	III	IV
		(reference value)			
C. Base case (with 10% probability for SDE-S[b])	0% (no SDE-P events)	(13.5 = 0)	+14.1	+9.2	−22.2
E. Base case, except 15% probability for SDE-P	15%	(19.6 = 0)	+19.8	+20.2	−1.6
F. Base case, except 30% probability for SDE-P	30%	(27.1 = 0)	+24.7	+30.5	+18.6
A. Standard base case	50%	(37.8 = 0)	+31.6	+44.1	+45.2

[a] Boxes denote best program choice.
[b] All cases assume a 10 percent probability for a security event.

These results suggest that even with smooth expectations for future oil price trajectories and no security events, possible uncertain outcomes for synfuel production costs and oil price trends combine to make some initial synthetic fuel production attractive relative to no production experience in the 1980s. A small program provides information benefits by clarifying the basic synfuel cost level as well as by providing some initial increment of technology improvement learning. These benefits appear to worthwhile even in a future where surge deployment events are absent. Although large initial production efforts appear costly and unattractive relative to more modest efforts, a large program becomes more attractive as the likelihood for a security event increases.

Varying Probabilities for Oil Price Jump Events. Table 6-6 summarizes the results for sensitivity cases that vary the probability for price events (*SDE-P*). Case C assumes that the probability of an *SDE-P* event is zero. (Security events occur with the base case probability of 10 percent for sensitivity cases, except as otherwise noted.) Program level II, or 250,000 barrels per day in the 1980s decade, is the preferred choice. If the probability of a price jump event in each decade is increased from 0 to 15 percent (Case *E*), Programs II and III yield the same level of benefits.[7] When the probability of price jump events is increased further to 30 percent (Case *F*), program level III becomes the best choice. For the standard base case (Case *A*) which assumes a 50 percent probability for *SDE-P* events, Program IV assumes a slight advantage over Program III. Higher probabilities for price jump events would further increase the relative attractiveness of program level IV.

Figure 6-5 summarizes these experiments with varying probabilities for oil price jump events. As one would expect for an insurance problem, relative program values are fairly sensitive to expectations for price jump events. Higher probabilities for oil price jump events not only increase the likelihood that the expanded and improved deployment capabilities obtained from initial production experience will be utilized, but also lower the cost of the insurance premium by increasing the benefits from production capacity installed prior to an event.

The relationship between recommended program size and the likelihood of major long-lasting price jump events is summarized in Table 6-7. The table shows that a moderate program (level III) is preferred across a broad range (15 to 50 percent) of probabilities for stylized oil price jump events (*SDE-P*). Low probabilities for *SDE-P* events favor a smaller program (II). Strong expectations for repeated price jump events favor the largest program level (IV) considered for this analysis.[8]

The decision analysis model can be used to assess the probabilities associated with any particular world view for oil price or security events. However, the sensitivity cases provide some intuitive feel for how relative program values change with varying expectations for such events. It appears that given base case specifications for other uncertainties, a moderate scale program is likely to be a good choice, unless one thinks that oil price jump events or security events are very likely.

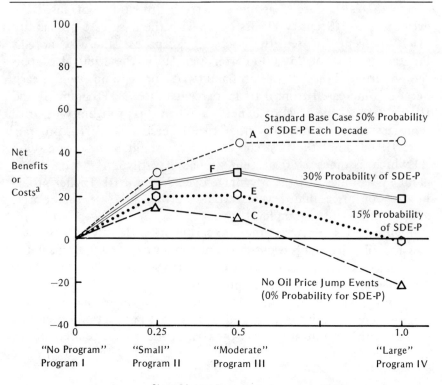

Figure 6-5. Sensitivity Cases Varying Likelihood of Oil Price Jump Events (SDE-P's)*

Size of Initial Production Program in 1980s (MMBD, Coal-Based Synfuels)

[a]Net Program Benefits Relative to "No Program" Reference Value (NPV in Billions of 1980$ at 5% Discount Rate)
*Probability of security event (SDE-S) 10% in each of these cases.

Exploring the Importance of Various Sources of Benefits

Sensitivity experiments can be used to explore the relative contributions from various sources of program benefits. As shown in Figure 4-7, different sources of benefits bear differing relationships to program size. Information benefits can be realized via an initial production effort comprising only a few projects. Learning benefits do not increase strongly with increasing program size. Infrastructure

Table 6-7. Probability Ranges for Price Jump Events for Which Particular Program Sizes are Recommended.

Label	Program Level 1980s Production	SDE-P Probability Range in Which Particular Program Level is Optimal
	(MMBD)	
II	0.25	0–15%
III	0.50	15–50%
IV	1.0	Greater than 50%

development, however, may exhibit a "critical mass" relationship where larger programs yield much greater deployability benefits.

Moreover, there are interdependencies between various sources of benefits. Enhanced deployment capability allows the cost savings from learning to accrue over a large volume of synfuel production (see Figure 4-8). Learning lowers costs for synfuel production, making larger deployment capabilities more likely to be utilized. These contrasting relationships and interdependent effects complicate assessments of the program size likely to provide the greatest level of benefits relative to costs. The sensitivity experiments described below may provide some insight.

A major design issue for initial synthetic fuel production programs is whether the additional deployability and learning benefits of larger initial production efforts are worth either the added near-term costs from factor cost inflation or the risks of large production subsidies. Aside from any direct production benefits (which occur when oil prices exceed synfuel production costs for initial plants), the major advantage of larger initial production programs is increased infrastructure development. Larger programs "buy" substantially more infrastructure development than smaller programs.

Table 6-8 and Figure 6-6 present the results of several sensitivity experiments designed to illustrate the relative importance of various sources of program benefits. (The standard base case run (Case A) is repeated in the table and figure for reference.)

"Pure" Learning Effect. Sensitivity Case G shows the effect of eliminating deployability benefits and examining the "pure" learning effect. This is accomplished by making the deployability increase

Table 6-8. Sensitivity Cases Illustrating Relationships Among Sources of Program Benefits (net expected program benefits relative to "no program" reference case; billions of 1980$ net present value at 5 percent discount rate).[a]

		Program Level			
Sensitivity Case		I	II	III	IV
		(Reference Value)			
A.	Standard base case	(37.8 = 0)	+31.6	+44.1	+45.2
G.	"Pure" learning effect (DIF = 0; BDC = 50[b])	(69.7 = 0)	+37.6	+41.1	+26.0
H.	"Pure" deployability effect with no learning	(32.4 = 0)	+20.3	+29.2	+27.2
J.	"Pure" deployability effect with 8% learning for all programs	(37.6 = 0)	+32.7	+43.5	+43.9
K.	Low base deployment capability (BDC = 10)	(19.3 = 0)	+25.8	+40.5	+45.5
L.	High base deployment capability (BDC = 40)	(66.9 = 0)	+37.3	+47.5	+43.3
M.	All programs with same expectations for benefit outcomes	(37.4 = 0)	+37.0	+44.5	+40.6
N.	Direct production benefits only: no learning benefits and no deployability effects	(51.6 = 0)	+19.1	+21.6	+8.0

[a]Boxes denote best program choice.
[b]BDC = 50 corresponds to 2.5 MMBD; BDC = 10 corresponds to 0.5 MMBD.

Figure 6-6. Sensitivity Cases Separating Learning and Deployability Effects

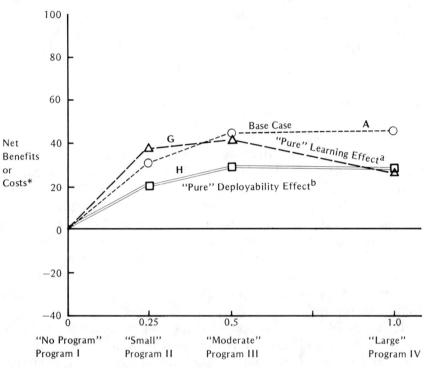

*Net Program Benefits Relative to "No Program" Reference Value (NPV in Billions of 1980$ at 5% Discount Rate)

[a] "Pure" learning effect eliminates infrastructure development effects by setting the linkage factor DIF equal to zero and making basic deployment capability 2.5 MMBD per decade.

[b] "Pure" deployability effect assumes no learning for any programs.

factor equal to zero and increasing the basic deployment capability from 1.0 MMBD per decade to 2.5 MMBD per decade. Consequently, the level of deployment possible for the 1990s and 2000s production expansion decisions is both large and completely independent of previous production levels. Since a major value of larger programs is to build expanded infrastrucutre capabilities, one would expect the larger program to look less attractive relative to smaller programs when deployability effects were eliminated.

This expectation is confirmed by the results of sensitivity Case G. The preferred program choice shifts from indifference between Programs III and IV in the base case (A) to a slight preference for Program III over Program II in the pure learning effect case (G). Even if price jump events occur with 50 percent probability, a large program (IV) looks unattractive relative to more modest efforts (Programs III and II) if deployability effects are unimportant.[9,10]

"Pure" Deployability Effects. Sensitivity Case H examines the "pure" deployability effect by setting learning benefits equal to zero for all programs. With learning effects absent, one would expect pure deployability effects to favor larger programs. The results of sensitivity Case H indicate that Program III is slightly preferred to Program IV. Thus, with learning effects eliminated, the added benefits of the additional deployment capability purchased via Program IV do not offset the added costs for the larger program. Case H results also suggest that the elimination of learning effects reduces net expected synfuel benefits by about one-third, or by $10 to $20 billion net present value.

It is possible that the "pure" deployability benefits would be larger if learning effects reduced the cost of future synfuel deployment and increased the value of expanded deployment capabilities. Sensitivity Case J examines this possibility by setting learning equal to 8 percent for all initial production programs (except the "no program" choice). Case J results show that the size net program benefits return to levels similar to those observed for the base case. Even with learning effects restored, however, a large program (IV) acquires no strong advantage over the moderate program (III), despite the central focus on deployability benefits for sensitivity Case J.

Assumptions for Base Deployment Capability. Sensitivity Cases K and L explore the importance of the deployability benefits further. The base deployment capability (BDC) represents the amount of future production expansion that is possible, irrespective of previous production levels. If the base deployment capability is relatively large, deployability benefits are less important. Conversely, if BDC is small, the importance of infrastructure effects is accented. Figure 6-7

Figure 6-7. Varying Specifications for Base Deployment Capability

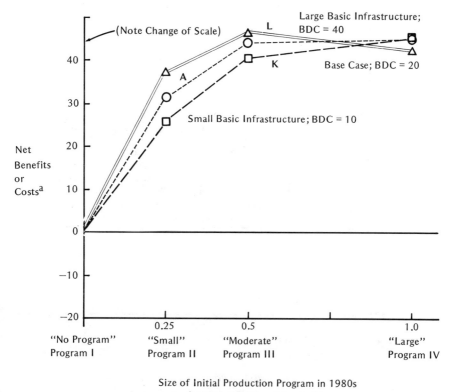

Size of Initial Production Program in 1980s
(MMBD, Coal-Based Synfuels)

Net Program Benefits Relative to "No Program" Reference Value
(NPV in Billions of 1980$ at 5% Discount Rate)

displays results of sensitivity cases that vary the assumption for base deployment capability from ten plants or 0.5 MMBD per decade (Case K) to forty plants or 2.0 MMBD per decade (Case L). (The standard base case (A) assumes a BDC of twenty plants or 1.0 MMBD per decade.) As expected, a small value for base deployment capability increases the relative importance of deployability effects and attractiveness of larger initial production efforts. However, if BDC is as large as 2.0 MMBD, the relative value of building additional strategic deployment capability via larger initial production efforts is substantially reduced.[11]

Figure 6-8. Sensitivity Cases for Alternative Descriptions of Program Benefit Relationships

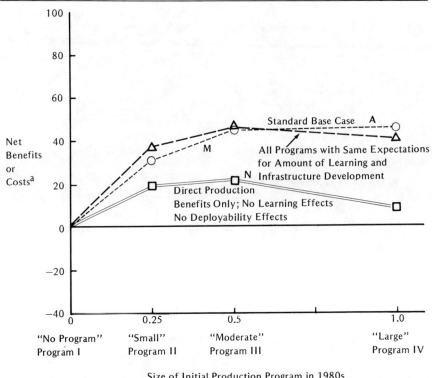

^{a}Net Program Benefits Relative to "No Program" Reference Value (NPV in Billions of 1980$ at 5% Discount Rate)

Changing Descriptions for Program Benefits. Sensitivity Cases M and N (displayed in Figure 6-8) further illustrate the effect of changing descriptions for program benefits. The base case (A) assumes that larger initial production efforts have increasing probabilities for larger benefit outcomes. In Case M, probabilities for various program benefit outcomes are assumed to be the same across all three program levels. Benefit expectations are constant or flat for all three programs.[12] As expected, these changed specifications for program benefit relationships increase the relative net benefits of Program II and decrease the net benefits of Program IV. Program level III emerges as the best choice, with net benefits which are 20 and 10 percent greater than Programs II and IV, respectively.

Sensitivity Case N examines the expected benefits for the direct value of the energy supplied by initial production facilities. These "direct production benefits" exclude benefits from future deployment of synfuels production. This is accomplished by simultaneously eliminating learning cost reductions and making 1990s and 2000–2010 deployment choices independent of levels of previous production. Thus, the estimates of net benefits for Case N reflect only the direct expected value of initial production volumes and the value of information about synfuel production costs that may improve some 1990 deployment decisions. The elimination of both learning and deployability effects reduces net benefits substantially from those estimated in the base case. Comparing Case N to the base case (A), one can see that learning and deployability effects combine to more than double the benefits of the base case as compared to a case (N) that reflects only the direct physical value of the energy supplied by initial synfuel production projects (and the value of information). For the direct value of initial production volumes only, program level III has a slight advantage over program level II, but Programs II and III are both much better than Program IV.

Summary. These sensitivity case results suggest that a moderate program (III) is the best choice across many specifications for program benefit relationships. When benefits are limited to learning effects (Case G), Program III has a slight advantage (9 percent) over a small program (II) and a substantial advantage (58 percent) over a large program (IV). When benefits are limited to deployability effects (Cases H or J), Programs III and IV yield similar benefits. For base case probabilities of surge deployment events, a small program (II) is substantially less attractive when deployability effects are important. Finally, when both learning and deployability effects are eliminated in order to examine the direct value of the Btus from initial production projects in isolation (Case N), Program III is slightly better than Program II and much better than Program IV.

How Factor Cost Inflation Affects Recommended Choices

Each of the previous sensitivity cases (G through N) examined changing descriptions for program benefits as a function of program size. During these comparisons, a major source of added costs for larger initial production programs—factor cost inflation—was held constant. This section reviews sensitivity cases that illustrate how appropriate program choices vary with the specifications for the amount of factor cost inflation.

Overall, the sensitivity experiments show that net program benefits vary greatly with assumptions regarding factor cost inflation. This sensitivity to assumptions for factor cost inflation combines with the relative uncertainty of this important parameter to recommend special attention to cost inflation effects as a key concern for program design. Since the sources and magnitude of the additional costs are not well understood, factor cost inflation is a priority issue for further study in the assessment and design of initial synfuel production efforts.

Table 6-9 summarizes the results of several sensitivity cases that vary the description of factor cost inflation from base case specifications. Figure 6-9 displays the corresponding net benefits curves.

Results with No Factor Cost Inflation. Sensitivity Case O shows the effect of completely eliminating the factor cost inflation from the analysis. The zero value reference point for the "no program" option (Program I) increases from $37.8 billion in the base case (A) to $76.3 billion in Case O. The relative net benefits of Programs III and IV increase substantially, with Program IV showing an increase of over $50 billion as compared to the base case. With factor cost inflation suppressed, the largest program level examined becomes the preferred program choice.[13]

Varying the Amount of Factor Cost Inflation. Preferred program choices are very sensitive to changes in specifications for factor cost inflation. For example, Case P1 shows the results of adding 5 percent to base case specifications for factor cost inflation for both Programs III (10 percent becomes 15 percent) and IV (25 percent becomes 30 percent). This modest change reduces the net benefits of program levels III and IV by about $5 billion and $11 billion, or 12 percent and 24 percent, respectively. The appropriate program choice shifts from a 2 percent advantage for Program IV over Program III, to a 12 percent advantage for the moderate program level (III).

Larger changes in the amount of factor cost inflation have even more striking effects. For example, in Case P2, the amount of factor cost inflation is doubled from base case assumptions (to become 20

Table 6-9. Sensitivity Cases for Factor Cost Inflation (net expected program benefits relative to "no program" reference case; billions of 1980$ net present value at 5 percent discount rate).[a]

	Sensitivity Case	Program Level			
		I	II	III	IV
A.	Standard base case[b]	(Reference Value) (37.8 = 0)	+31.6	+44.1	+45.2
O.	Base case with no factor cost inflation	(76.3 = 0)	+29.2	+54.9	+99.3
P1.	Slightly higher factor cost inflation[c]	(32.3 = 0)	+31.5	+38.7	+34.5
P2.	Double factor cost inflation[d]	(26.4 = 0)	+27.9	+29.6	−7.6
Q.	Factor cost inflation in 1980s decade only	(52.6 = 0)	+52.8	+69.6	+78.0

[a] Boxes denote best program choice.

[b] Base case factor cost inflation assumptions: Program II (+0%); Program III (+10%); Program IV (+25%).

[c] Base case with 5% additional factor cost inflation for Programs III and IV: Program III (+15%); Program IV (+30%).

[d] Base case with twice as much factor cost inflation for Programs III and IV: Program III (+20%); Program IV (+50%).

percent for Program III and 50 percent for Program IV). Results for Case P2 show that Program III is the best choice, with net benefits of $30 billion (NPV) giving it a very slight advantage over the small program's (II) benefits of $28 billion. Case P2 results show that the added costs from high levels of factor cost inflation can quickly undermine any added deployability advantages of the large program (IV); net benefits drop from $45 billion (NPV) in the base case (A) to negative net benefits (net costs) of $8 billion (Case P2).

Figure 6-9. Sensitivity Cases Varying Amounts of Factor Cost Inflation

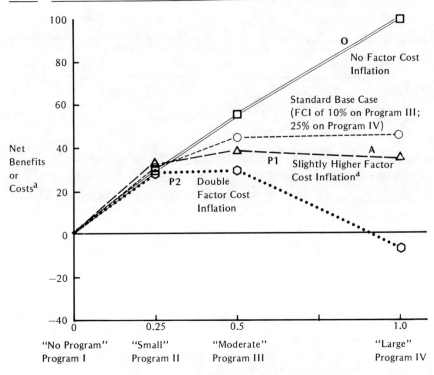

[a] Net Program Benefits Relative to "No Program" Reference Value (NPV in Billions of 1980$ at 5% Discount Rate)

The striking difference in expected benefits for a large program in Cases *P*1 and *P*2 affirms the importance of understanding and avoiding excessive factor cost inflation for initial synfuel production efforts. Factor cost inflation is poorly understood and inherently difficult to estimate. However, many experts have strong expectations for substantial factor cost inflation, if existing deployment capabilities are overextended. As observed by Cameron Engineers in their report to a U.S. Senate committee, "costs during an all-out crash program are likely to increase by 50% or more" (Cameron engineers 1979: 27). The net benefits of larger initial production programs are especially sensitive to factor cost inflation specifications, because any added costs are experienced over a larger production volume. High levels of factor cost inflation can easily offset any learning and deployability advantages of larger initial production programs.

Estimating Crossover Levels of Factor Cost Inflation. The standard base case (*A*) assumes that factor cost inflation occurs for larger deployment levels in each of the three decades examined in the decision analysis. Sensitivity Case *Q* assumes that factor cost inflation occurs in the initial deployment decade only. Maximum deployment increments in the decades following initial production do not incur factor cost inflation. Figure 6-10 compares results of sensitivity Case *Q* with the base case. Although the net benefits of all programs have increased noticeably for Case *Q*, the general shape of the net benefits curve is similar. Just as for Case *A* results, the curve for Case *Q* shows substantial benefits to a small initial production program (Program II), significant additional benefits from expanding to program level III (despite 10 percent factor cost inflation), and only modest additional benefits from the largest program (IV).[14] This similarity of relative program values between Cases *A* and *Q* provides the basis for some useful hand calculations regarding acceptable amounts of factor cost inflation as described below.

If factor cost inflation occurs in only one decade, the costs due to factor cost inflation for larger programs can be calculated by hand separately from the decision analysis model. For example, suppose factor cost inflation increases the costs for a $50 per barrel basic synfuel cost level by 10 percent or $5 dollars per barrel. The net present value (NPV) of the $5 per barrel increase can be calculated with simple arithmetic to be about $0.86 billion for each representative initial production plant.[15] The amount of expected cost increases due to factor inflation depends on the basic synfuel cost level and the percentage amount of factor cost inflation. For the base case specifications of synfuel cost outcomes and probabilities, each ten percentage points of factor cost inflation results in added per barrel costs that total to additional costs of $0.9 billion (NPV) for each representative project that is part of an initial production program.[16]

Table 6-10 displays estimates of the expected amount of reduction in net program benefits due to factor cost inflation for various

Figure 6-10. Similar Shape Net Benefits Curves for Alternative Modeling of Factor Cost Inflation

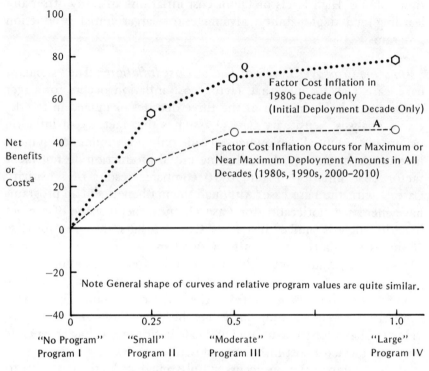

Size of Initial Production Program in 1980s
(MMBD, Coal-Based Synfuels)

[a] Net Program Benefits Relative to "No Program" Reference Value
(NPV in Billions of 1980$ at 5% Discount Rate)

---

size initial production programs. The amount of "benefit reduction" increases for larger initial production efforts, because the added costs are experienced over a larger production volume. Though factor cost inflation may be caused primarily by the last few plants in a program, the added costs affect all of the plants built. The amount of added costs or benefit reduction also increases in direct proportion to the level of factor cost inflation assumed. For program level IV (twenty plants), 10 percent factor cost inflation reduces benefits by $18 billion (NPV), 20 percent reduces net benefits by $36 billion, and so forth.

Table 6-10. Amounts of Benefit Reduction Due to Factor Cost Inflation (total expected additional costs for various program levels; expressed in billions of 1980$, net present value at 5 percent discount rate).

Program Level	Assumed Amount of Factor Cost Inflation (as percentage increase from basic synfuel cost outcome)		
	5%	10%	30%
Cost Per Nominal Plant	0.45	0.90	2.7
II: (5 plants)	2.2	4.5	13.5
III: (10 plants)	4.5	9.0	27.0
IV: (20 plants)	9.0	18.0	54.0

Such estimates of the amount that factor cost inflation reduces benefits—when combined with decision analysis model runs excluding factor cost inflation—allow hand-calculated sensitivity experiments with varying assumptions about the amounts of factor cost inflation. For example, if Program IV had net benefits of $68 billion with no factor cost inflation, 10 percent factor cost inflation would reduce benefits by $18 billion (from Table 6-10) to yield net benefits of $50 billion.

Figure 6-11 illustrates how the benefit reduction estimates of Table 6-10 may be used to estimate the amount of factor cost inflation which would make different size programs yield the same level of benefits. In Figure 6-11, results for the sensitivity case that excludes factor cost inflation (Case O, Table 6-9) are graphed with net benefits on the vertical axis and size of initial production program on the horizontal axis. The difference in benefits between program levels III and IV is $44 billion, as indicated by the brackets. Referring to Table 6-10, we see that if each ten percentage points of factor cost inflation reduces the net benefits of Program IV by $18 billion, about 25 percent factor cost inflation will reduce net benefits from Program IV to the same level as that expected for Program III. Similarly, for the comparison of program levels II and III, the $26 billion advantage for Program III will be completely offset if program level III experiences 29 percent factor cost inflation. By parallel calculations, the net benefits of Program IV will be reduced to the same value as those expected for Program II if Program IV experiences 39 percent factor cost inflation.[17]

Figure 6-11. Illustration of "Crossover" Amounts of Factor Cost Inflation Which Equilize Benefits of Different-Size Programs.

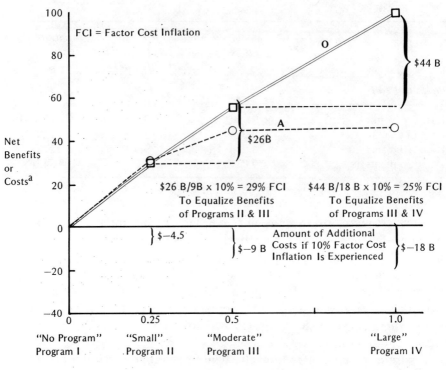

Such calculations are useful because they give a feel for the expectations one must have about magnitude of factor cost inflation in order for the appropriate program choice to shift or "crossover" from one program to another. Table 6-11 displays estimates of the "crossover" levels of factor cost inflation that would reduce the benefits estimated for larger programs without any factor cost inflation to the same level as the benefits estimated for smaller programs.[18] Estimates are presented for selected sensitivity cases analogous to sensitivity cases previous described.

Table 6-11. Crossover Levels of Factor Cost Inflation (percentage factor cost inflation that will make the expected benefits of larger program equal to those of smaller program).

Sensitivity Case[a]	A IV with III	B IV with II	C III with II	Ratio of Column B to Column C
O/E. Standard base case without factor cost inflation	25%	39%	29%	1.34
B/E. No events (SDE-P or S) smooth future	0.8%	2.6%	3.4%	0.76
C/E. No SDE-P 10% SDE-S	4.3%	8.1%	7.4%	1.09
E/E. 15% probability of SDE-P 10% probability of SDE-S	7.4%	13%	10%	1.30
F/E. 30% probability of SDE-P 10% probability of SDE-S	17%	27%	20%	1.35
G/E. "Pure" learning effect	11%	17%	13%	1.31
H/E. "Pure" deployability effect with no learning	22%	34%	23%	1.48
J/E. "Pure" deployability effect with 8% learning for all programs	24%	37%	26%	1.42
K/E. Low base deployment capability	27%	44%	34%	1.29
N/E. Direct value of energy production only; No learning or deployability effects	11%	17%	12%	1.42

[a] The letter designations in this table refer to sensitivity cases previously described. For example, case F/E refers to the base case with 30 percent probability for price jump events from Table 6-6. The designation "/E" indicates it is a model run analogous to the previous sensitivity case, except that factor cost inflation eliminated in order to allow hand calculation of the crossover levels of factor cost inflation.

Estimates from Table 6-11 illustrate and quantify the significant influence of factor cost inflation for determining optimal program choices. For the base case assumption of 50 percent probability for price jump events, relatively high levels of factor cost inflation (on the order of 20 to 30 percent) can be accepted before the benefits of larger programs are reduced to those of smaller programs. However, as noted in Chapter 3, the 1973-74 experience of hyperinflation in chemical plant construction costs suggests that factor cost inflation could readily exceed 30 percent. Moreover, for lower probabilities of surge deployment events (such as those represented by cases *B/E*, *E/E*, or *F/E*), relatively modest expectations for factor cost inflation could make a moderate program (III) better than the large program (IV), or even a small program (II) superior to both the moderate and large programs.

How Much Factor Cost Inflation Is Acceptable? The first five cases of Table 6-11 illustrate the strength of the direct tradeoff between the likelihood of surge deployment events and the amount of factor cost inflation which is "acceptable" before recommended program size preferences "crossover" from larger toward smaller efforts. Figure 6-12 displays the estimates from Table 6-11 with the crossover level of factor cost inflation (vertical axis) plotted against the probability for surge deployment events (horizontal axis). Curves *A*, *B*, and *C* correspond to the columns designated in Table 6-11. The steep slope of these curves indicates the strength of the tradeoff.

Figure 6-12 provides a graphic basis for some intuitive feel for the expected amount of factor cost inflation compared to the likelihood of insurance events for smaller programs to be preferred to larger programs, or vice versa. For combinations of factor cost inflation and the likelihood for events in the upper left section of the plot, smaller programs are likely to be better than larger ones. In the lower right region of the plot, larger programs are likely to be a better choice. Curves *A*, *B*, and *C* divide the regions for comparisons of particular pairs of programs (IV with III, IV with II, and III with II).

Tradeoffs Between Infrastructure Development and Factor Cost Inflation. The last five cases in Table 6-11 illustrate the strong influence of infrastructure development effects for determining the amount of factor cost inflation that is acceptable for larger programs in order to build infrastructure for enhanced deployment capability

THE DECISION ANALYSIS RESULTS 201

Figure 6-12. Variation in "Acceptable" Levels of Factor Cost Inflation (FCI) with Likelihood for Surge Deployment Events[a]

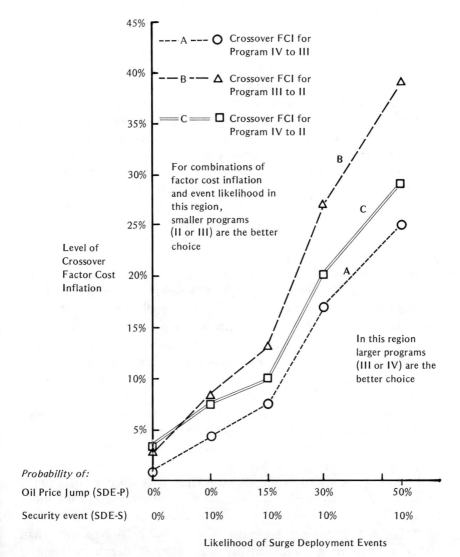

[a]For an example of how to use this plot see Note 19.

Comparing Case G/E with Cases H/E and J/E, the acceptable level of crossover factor cost inflation more than doubles in going from a situation where deployability effects are not important (Case G/E) to cases (H/E or J/E) where infrastructure development is a major concern.[20]

The Advantage of a Moderate Program if Factor Cost Inflation Is Likely. Acceptable levels of factor cost inflation depend strongly on expectations for surge deployment events and the importance of infrastructure development. However, in all cases examined, crossover levels are by no means so high as to be implausible. This observation holds with particular force when one compares the difference between the levels of factor cost inflation that would reduce the benefits of a moderate (III) or large (IV) program to equal the benefits of a small program (II). For example, in Case E/E, 10 percent factor cost inflation would reduce the benefits of Program III to equal those of Program II. (Factor cost inflation in excess of this 10 percent crossover level would make Program II better than Program III.) However, only an additional three percentage points (or 13 percent total) of factor cost inflation for a large program (IV)—which is twice the size of Program III—would reduce the benefits for the large program (IV) to those of the small program (II). Since the amount of factor cost inflation is likely to increase in at least direct proportion to program size or probably even increase more strongly, a given percentage of factor cost inflation for Program III implies at least twice that amount for Program IV. The column on the far right of Table 6-11 shows the ratio of crossover levels of factor cost inflation for a moderate program (III) with a small program (II), and the large program (IV) with a small program.

If factor cost inflation increases in direct proportion to program size, the ratio would be expected to be at least two, but the ratio in Table 6-11 is always less than two. Thus, if factor cost inflation is likely to be serious enough to be near the crossover level for a moderate program compared to a small program (shown in Column C of Table 6-11), a large program is very likely to be worse than either a small or a moderate program. If a moderate program is likely to experience little or no factor cost inflation, Column A of Table 6-11 defines the crossover amount of factor cost inflation for a large program compared to Program III. Column A values show that relatively modest amounts of factor cost inflation for Program IV could make a moderate effort yield greater benefits than a large program. Beyond some point, increasing program size is very likely to be counter-productive.

Summary Taken together, these observations from Table 6-11 suggest that, if limited infrastructure capabilities make factor cost inflation a serious concern, a moderate program (III) may be better than a large program (IV) for a wide range of views of the world or expectations for the amount of factor cost inflation. Although less likely, some levels of factor cost inflation could make even make a small program the preferred choice. On the other hand, if factor cost inflation is likely to be small or zero, larger initial production efforts become increasingly attractive. If we are sure that enhanced deployment capability is important, either because surge deployment is fairly likely or because existing deployment capability is quite limited, it may be appropriate to tolerate a modest amount of factor cost inflation in order to build the industry infrastructure.

The decision analysis answers to the "How large?" question are very sensitive to expectations for factor cost inflation. Clearly, this poorly understood, but important effect merits explicit, perhaps priority, consideration for the assessment, design, and implementation of initial synthetic fuel production efforts.

Alternative Descriptions for Surge Deployment Events

The basic notion of a rapid increase in the market price or security costs of imported oil may be readily imagined. However, description and modeling for such events is more problematic. The decision analysis model necessarily assumes a standard stylized description of both price jump events (*SDE-P*) and security events (*SDE-S*). Decision analysis results depend on how such events are modeled and the probabilities assumed for the events. This section presents results of sensitivity experiments that examine the effect of alternative descriptions for surge deployment events.

Alternative Descriptions for Price Jump Events. The standard description of a price jump event (*SDE-P*) is a doubling (100 percent

Table 6-12. Sensitivity Cases for Alternative Descriptions of Price Jump Events (net program benefits relative to "No program" reference case; billions of 1980$ net present value at 5 percent discount rate).[a]

Sensitivity Case		Program Level			
Variation in Amount of Price Jump		I ("No Program" Reference Value)	II	III	IV
Case	Percentage Jump				
C.	0%[b]	(13.5 = 0)	+14.1	+9.2	−22.2
R.	50%	(24.2 = 0)	+23.0	+26.5	+10.8
S.	75%	(30.7 = 0)	+27.2	+35.2	+27.8
A.	100%[c]	(37.8 = 0)	+31.6	+44.1	+45.2
T.	125%	(45.9 = 0)	+35.5	+52.5	+62.5
Variation in Length of Decline Period					
Case	Years				
U.	3	(16.2 = 0)	+17.7	+16.4	−8.3
V.	10	(28.5 = 0)	+26.0	+34.3	+26.7
A.	15[c]	(37.8 = 0)	+31.6	+44.1	+45.2
W.	20	(48.1 = 0)	+34.6	+50.9	+58.9

[a] Boxes denote best program choice.
[b] Case C reflects no SDE-P events occurring. However, all cases in the table assume a 10 percent likelihood for security events.
[c] Value for standard base case specificiations.

increase) in the market price of oil followed by a linear decline over fifteen years back to some underlying long-run trend. Two factors—the percentage oil price jump (OPJ) and the length of the decline period (LDP)—are central to the description of a price jump event. Table 6-12 summarizes the results of sensitivity cases in which these two parameters are varied across a range of values. For all of these cases, the probability of an SDE-P event is the standard base case specification of 50 percent in each decade. Only the modeling for the events changes.

Amount of Oil Price Jump. Cases varying the percentage oil price jump (OPJ) from 50 to 125 percent are shown in the top half of Table 6-12. Figure 6-13 displays the results graphically. As expected, increases in the amount of oil price jump over the standard 100 percent favor larger programs relative to smaller programs. Less severe oil price jumps act in the opposite direction. Net benefits estimated for program level IV are more sensitive to the specification of the percentage oil price jump than Programs III or II. In general, for oil price jumps less than 100 percent, Program III is better than Program IV.

Length of Oil Price Jump Events. Cases varying the length of decline period (LDP) are presented in the lower part of Table 6-12. Again, relative program benefits change in intuitively expected directions. When the length of an event is relatively short (i.e., LDP is small), smaller initial production programs appear better than larger programs. Longer events increase the attractiveness of larger initial production efforts.[21] Case U results reaffirm the qualitative point that, since synfuels take a long time to deploy, large deployment efforts should not be based on relatively short term (three-year) price jump events, even if those events are very likely to occur.

Alternative Descriptions of Security Events. Security events may also be modeled differently than the standard description assumed for the base case. Figure 6-15 illustrates the sensitivity of model results to the description for a security (*SDE-S*) event.

Cases B, C, and D (from Table 6-5) repeat results of assuming 0, 10, and 30 percent probabilities for a standard *SDE-S* event. Case X shows the effect of a more severe description for a security event where the oil import premium increases from $5 per barrel to $40 per barrel (rather than only $20 per barrel as assumed for the standard description for a security event). As expected, larger programs are favored and the best decision shifts from Program II to Program III. Sensitivity Case Y reflects the same, more severe description of an *SDE-S* event combined with standard oil price jump events occurring with the base case probability of 50 percent in each decade. In Case Y as compared to the base case (A), a more severe security event increases the advantage of Program IV as compared to Program III from 2 to 19 percent. In general, the relative program values are not extremely sensitive to the modeling for security events.

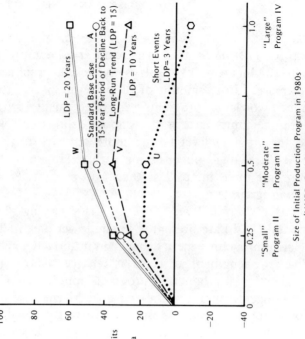

Figure 6-13. Sensitivity Cases Varying Amount of Oil Price Jump

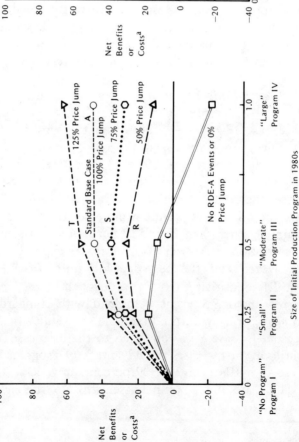

Figure 6-14. Sensitivity Cases Varying Length of Price Jump Event (Length of Decline Period Back to Long Run Trend "LDP")

Figure 6-15. Sensitivity Cases for Modeling of Security Event (SDE-S)

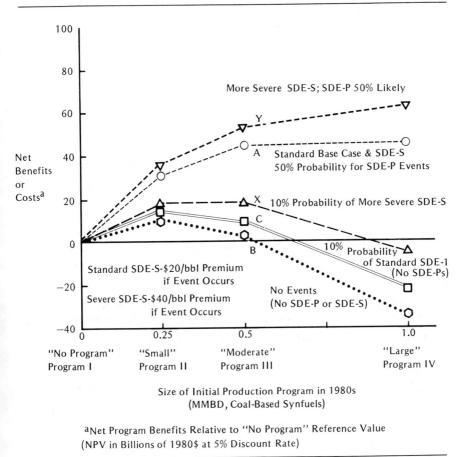

Size of Initial Production Program in 1980s
(MMBD, Coal-Based Synfuels)

[a]Net Program Benefits Relative to "No Program" Reference Value
(NPV in Billions of 1980$ at 5% Discount Rate)

Sensitivity Cases for the Value of Infrastructure Development

The major insurance advantage of larger initial production programs is their contribution to infrastructure development. Two major elements of the decision analysis specifications bear importantly on the value of larger initial production efforts for buying expanded surge deployment insurance capability. First, the synfuel production limit constrains maximum deployment production levels irrespective of infrastructure development and thus could make expanded deployment capability unusable. Second, the rate at which practical production experience builds infrastructure to increase future deployment capability is not well understood. The appropriate range for the linkage factor (DIF) which quantifies the rate of infrastructure development is uncertain. This section briefly reviews some sensitivity cases that vary from the base case values specified for these two uncertain parameters.

Synfuel Production Limit. The synfuel production limit (SPL) reduces the benefits of increasing deployment capability. For example, a maximum deployment expansion capability of 2.0 MMBD per decade in decision year 2000 would be of little value if the synfuel production limit were 2.5 MMBD and the existing production level in the year 2000 were 2.0 MMBD.

Table 6-13 presents the results for sensitivity Case Z where the synfuel production limit (SPL) is increased to 8.0 MMBD from the base case specifications for the uncertain range of SPL outcomes. (The base case range has an expected value of 5.25 MMBD.) If we are sure that the synfuel production limit is high, larger initial production programs have markedly increased benefits. For example, with base case SPL assumptions (Case *A*), Program IV is only two percent better than Program III. However, with high synfuel production limits (Case *Z*), the advantage of program level IV increases to 29 percent.[22]

Case *AA* displays results if a low outcome for the synfuel production limit is certain to occur. The low outcome limits synfuel production levels to no more than 2.5 million barrels per day through 2010. A large initial production program becomes less attractive than a small program. With a low synfuel production limit, the additional deployment capability purchased via a larger program either is unnecessary (for example, if oil imports have been completely eliminated) or cannot be fully used (if constraints other than infrastructure capabilities limit the role for synfuels).

Figure 6-16 illustrates the sensitivity of decision analysis results to specifications for synfuel production limits. The net benefits of larger programs are particularly sensitive to expectations for synfuel production limits. For example, comparing Cases *A* and *Z* (Table 6-13), the increase in net benefits for Programs II, III, and IV obtained from relaxing the synfuel production limit are 5, 16, and 46 percent, respectively. Comparing the base case with Case *AA*, a

Table 6-13. Sensitivity Cases Illustrating the Value of Infrastructure Development (net program benefits relative to "No program" reference case; billions of 1980$ net present value at 5 percent discount rate.)[a]

Sensitivity Case		Program Level			
Label	Description	I	II	III	IV
		("No Program" Reference Value)			
A.	Standard base case with SPL expected value of 5.25 MMBD	(37.8 = 0)	+31.6	+44.1	+45.2
Z.	High synfuel production limit (8 MMBD)	(38.6 = 0)	+33.3	+51.2	+66.1
AA.	Low synfuel production limit (2.5 MMBD)	(36.8 = 0)	+26.3	+31.0	+21.7
BB.	Stronger infrastructure development effect[b]	(46.0 = 0)	+36.0	+55.3	+64.4

[a] Boxes denote best program choice.
[b] The stronger infrastructure development effect is achieved by examining broader range of DIF outcomes. The range examined in Case BB is 0.5, 1.5, and 3.0. The range of DIF outcomes for base case was 0.3, 0.8, 1.4. The probabilities for various outcomes with different initial program choices are the same for Cases A and BB.

low synfuel production limit reduces the net benefits of Programs II, III, and IV by 17, 30, and 52 percent, respectively.

These results suggest that one's expectations for those factors that could limit the maximum mid-term market role for coal-based synfuels—such as logistical constraints on coal production, elimination of oil imports via other domestic supply and demand responses, and so forth—should affect near-term preferences for the size of an initial production effort. For example, if optimism is high for heavy oil or natural gas production or even oil shale, the appropriate size for initial production efforts with coal-based synfuels may be smaller.

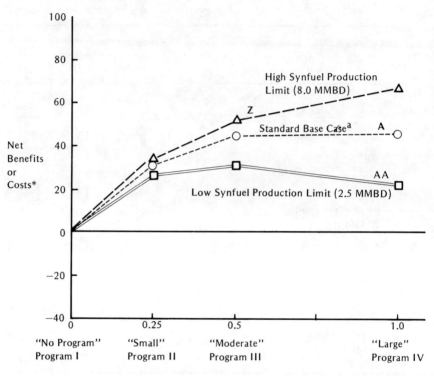

Figure 6-16. Effect of Synfuel Production Limit Outcome on Value of Larger Programs

*Net Program Benefits Relative to "No Program" Reference Value (NPV in Billions of 1980$ at 5% Discount Rate)

[a]Expectation for Synfuel Production Limits Uncertain; 3 Outcomes—2.5, 5.0, 8.0 MMBD—with Expected Value of 5.25 MMBD.

It is less worthwhile to invest in substantial infrastructure development if coal-based synfuels are likely to have a small role in the energy supply mix at the energy price levels required to make them generally economic. Accordingly, the potential mid-term (through 2010) market role for the coal-based synfuels option is relevant to the 1980s decision on "How large?" an insurance investment should be made.

The size of the role for synthetic fuels in the mid-term energy supply mix depends on many uncertainties, including other domestic energy supply and demand responses as well as economic growth. However, many energy/economic models predict low U.S. oil import demand at the high energy price levels that would be associated with oil price jump events. Expectations for the maximum extent of the mid-term role for coal-based synfuels should be explicitly considered in assessments of the value of investments in infrastructure development and the scale of initial production efforts.

Strength of Infrastructure Development Effect. The process by which deployment capability expands is not well understood. Consequently, the deployability increase factor (DIF) is explicitly modeled as an uncertain outcome. Base case specifications examine outcomes for DIF ranging from 0.3 to 1.4; however, stronger linkages between production experience and expanding future deployment capability are plausible. With higher DIF values, initial productione efforts would have greater benefits since they could buy more deployment capability. Case *BB* in Table 6-13 displays results when the range of DIF outcomes reflects an infrastructure development effect roughly twice as strong as the range assumed for the base case (outcomes of 0.5, 1.5, and 3.0 rather than 0.3, 0.7, 1.4 for the base case).

As expected, a stronger infrastructure development effect markedly increases the relative advantage of larger initial production efforts. A moderate program (III) is 54 percent better than a small program in Case *BB* as compared to 29 percent for the base case (*A*); a large program is 14 percent better than Program II as compared to only 2 percent in the base case. Clearly, specifications for the highly speculative DIF parameter have a strong influence on relative program values. Improved understanding of infrastructure growth effects is important for assessing the insurance value of larger initial production efforts.

Are Free Market Incentives Alone Adequate?

Up to this point, assessments of the appropriate level for initial production efforts have been based on the expected benefits, costs, and preferences for the nation as a whole—a "social" accounting or objective function. No attempt has been made to separately identify benefits that could provide profits for private firms motivating them to proceed with some initial production projects in response to free market opportunities without any special government "program." However, private firms may be able to capture some of the benefits of initial synfuels production as profits. Private decisions in response to market incentives alone could lead to some initial synfuels production experience.

The merits of special government programmatic efforts depend on whether expected private-market choices alone will lead to the level and form of initial production experience that seems appropriate for the nation as a whole. If the optimal program for the nation differs significantly from the choices and levels of investment that private firms would be expected to make in response to market incentives alone, special government action or incentives (a "program") to promote additional private sector efforts may be warranted. However, if free market incentives are likely to lead to investment decisions and levels of production experience similar to the optimal choice for society, special government program efforts may be both unnecessary and inappropriate.

Benefits and appropriate choices for the nation as a whole may differ or diverge from those for individual private firms for several reasons. First, private firms cannot be expected to account for the oil import premium or the potential security events that might increase the premium. Private firms also may make decisions using a decision rule or objective function incorporating a different time value of money or a different attitude toward risk than the nation as a whole may apply.[23]

Secondly, only a portion of the benefits from initial production projects may be captured as revenue by the private firms whose pioneering projects produce the benefits. For example, the benefits of improved infrastructure expand options and reduce bottlenecks for the industry as a whole, but are paid for primarily by the firm whose project provided the experience that trained workers, engineers, and managers, expanded component suppliers, and developed the infrastructure. Economists apply the term "positive externalities" to the circumstances where benefits accross society exceed benefits that are captured and counted by private firms.

Thirdly, the private sector may have different expectations for the future or the potential profitability of some future scenarios. A "windfall events" tax would reduce the value to private firms of preparing to respond to potential surge deployment events.

Table 6-14 displays results of sensitivity experiments with the decision analysis model. These results illustrate quantitatively the effect of each of the three sources of divergence between expected private decisions on synfuel production levels and the levels that seem appropriate for the nation as a whole. These cases all assume the base case specifications for long-run oil prices, synfuel costs, and price jump events unless otherwise noted. The base case results (*A*) characterize the appropriate choice for the nation. Then expected private choices are examined for various descriptions of (1) the decision rules or objective function of private firms; (2) positive externalities for learning and deployability benefits; and (3) different expectations for future events or ability of private firms to reap higher returns during such events.

Different Decision Rules. The analysis of appropriate choices for the nation assumes a rate of time preference (social discount rate) of 5 percent in real terms. However, a private sector decisionmaker might evaluate the costs and benefits according to a real pre-tax discount rate. If we assume that corporate stockholders have the same time preference as society for consuming now as compared to investing for future consumption (i.e., the same 5 percent discount rate), the corresponding pre-tax discount rate is about 10.7 percent.[24] This higher discount rate is assumed to apply to the private sector synfuels project decision rule. The qualitative effect of the higher discount rate is to weight near-term costs more heavily than later year benefits. The higher discount rate also simulates the effect of higher rates of return or capital charge rates for relatively risky synfuel projects. The qualitative effect is to magnify costs relative to benefits. Less synfuel production would be expected.

Case *CC* of Table 6-14 illustrates this result quantitatively. Specifications are the same as the base case, except a 10 percent discount rate is used. The recommended choice shifts to Program III which has benefits 18 percent greater than Program IV (as compared to 2 percent less for the base case at a 5 percent discount rate).

In addition to a higher discount rate, private firms would not be expected to include the value of the oil import premium or potential security events in their synfuels decisions. Even if learning and infrastructure benefits were completely reflected in the assessments of private firms, the elimination of the oil import premium and security events would decrease the value of initial production projects

Table 6-14. Sensitivty Cases Illustrating Potential Divergence between Optimal Choice for the Nation and Expected Private Incentives and Choices (net program benefits relative to "no program" reference case; billions of 1980$ net present value at indicated discount rate).[a]

Sensitivity Case		Program Level (MMBD)			
		I	II	III	IV
		("No Program" Reference Case)	(0.25)	(0.5)	(1.0)
A.	Nation's choice[b]	(37.8 = 0)	+31.6	[+44.1]	+45.2
CC.	Nation's choice with 10% discount rate	(8.1 = 0)	+8.5	[+11.4]	+9.3
DD.	Private choice if all benefits are privately appropriable[c]	(3.2 = 0)	[+4.6]	+4.3	−12.9
EE.	Private choice if no benefits are privately appropriable[d]	(5.8 = 0)	[+3.6]	+2.0	−7.3
FF.	Private choice with only learning benefits appropriable[d]	(6.5 = 0)	[+6.2]	+4.8	−4.6
GG.	Private choice if windfall events tax is expected[f]	[(0.4 = 0)]	−1.2	−7.0	−24.7

[a] Boxes denote best program choice.
[b] Standard base case social benefit calculation (5 percent discount rate).
[c] Private objective function*; future learning and deployability effects included.
[d] Private objective function*; no benefits from learning or deployability effects; price jump events yield full increments to private revenues.
[e] Private objective junction*; price jump events yield full increments to private revenues (i.e., no windfall events tax); no deployability benefits; learning benefits only.
[f] Same as case *HH* above, except revenues in price jump events completely captured by windfall events tax (no additional private benefits if oil price jump event occurs).

*Private objective function denotes no oil import premium or *SDE-B* (security) events and a 10.7 percent discount rate to reflect effect of taxation or risk premia.

and the expected level of private investment. Case *DD* of Table 6–14 displays the results of assuming a "private objective function," defined as a 10.7 percent discount rate with no accounting for the oil import premium or security events that might increase the premium. Program II becomes the best choice for private firms, even if learning and deployability benefits are completely appropriable.

Positive Externalities. Many of the benefits of a pioneering synfuels project do not result in revenues for the private firm undertaking the project. Case *EE* illustrates results if private firms make choices that completely ignore the benefits of learning and deployability increase for future production expansion and consider only the expected payoff from the energy supplied from initial production projects.[25] Program level II is the best private choice.

Case *EE* results assume that both learning and infrastructure development effects are not considered by private decision makers. However, private firms may account for learning benefits from initial projects since know-how can be sold and protected to some extent by the patent and technology licensing system. Case *FF* assumes that private choices include learning benefits but not infrastructure development effects. This less limited view of the benefits available to pioneering private firms increases the relative attractiveness of larger initial production choices, but a small program remains the expected private choice.[26]

Expectations for Future Events or Their Potential Payoff. Individual private firms could have expectations for future events or for the payoff from initial production projects in such events that differ from the expectations of the nation as a whole. This is particularly true for potential jumps in the market price of oil.

Given equivalent information, there is no reason to expect private firms' judgments about the likelihood and description of potential price jump events to be different (or better or worse) than the expectations the government may apply for assessing appropriate choices for the nation. Government may, however, have different ranges and sources of information than the private sector, and thus, different expectations.

Even if the expectations for price jump events of private firms and the government were identical, recent experience with the oil price controls instituted in response to price jump events during the 1970s and political attitudes regarding energy industry profits might cause firms to doubt whether a corresponding increase in synfuel revenues would actually be realized. Special "windfall event" taxes or price controls could reduce the revenues that synfuel producers might otherwise receive if a price jump event occurred. The mere expectation that such windfall event taxes might be imposed reduces private firms' incentives for investing in initial synfuels projects as a hedge against such events.

For example, a 50 percent windfall event tax would be analogous to changing expectations for the amount of an oil price jump from 100 percent to 50 percent. If a private firm believed there was a 50-50 chance of either a 100 percent windfall event tax or of product price controls being imposed during an event, it would make decisions as if doubling oil prices leads to only a 50 percent increase in synfuels project revenues. Previous sensitivity cases (C, R, and S) have illustrated the strong effect that reducing the expected amount of a price jump has for reducing the preferred amount of initial production.

Case GG in Table 6-14 illustrates an extreme view of privately appropriable benefits and expectations for windfall event taxes. This case applies a private objective function with no learning or deployability benefits, just as for Case EE. It further assumes that a windfall event tax is certain to be imposed and capture any increase in revenues should an oil price jump event occur. This effectively causes private firms to assess synfuel investment decisions as if the probability of a price jump event were zero, even if such events were likely to occur. Decision analysis results for this extreme description of a private firm's incentives indicate that private firms' preferred choice would be the "no program" option. Private investors would choose not to produce any coal-based synthetic fuels in the 1980s decade.

Summarizing the Potential Divergence. The qualitative differences in decision rules, appropriable benefits, and expectations for revenues in future scenarios suggest that private choices in response to market incentives alone might well result in less initial synfuels production than seems appropriate for the nation as a whole. However, the question of how much less bears importantly on the extent to which any special government program to promote initial synfuels production may be necessary or warranted. The net benefits of any government program should be assessed relative to a reference case of what would happen without special incentives.

Unfortunately, it is quite difficult to quantitatively assess this market-only reference case. The stylized cases of Table 6-14 are merely suggestive of what private firms might choose to do given alternative descriptions of their decision rules and time preferences, the benefits figuring in private firms' investment assessments, and expectations for windfall event taxes or price controls. These factors are difficult to determine and could vary significantly from one firm to another. Some firms may consider learning and infrastructure development effects; others may not. Some firms may think windfall event taxes are quite likely, while others think them unlikely.

Table 6-14 provides a rough quantitative feel for the divergence between the nation's choice (for base case specifications) and expected private choices. In general, when the nation would choose a moderate or large program (Case *A*), the combined choices of private firms would be expected to lead to only a small program (Cases *DD, EE, FF*). If windfall event taxes on price controls were considered likely (Case *GG*), private firms might choose not to produce any coal-based synfuels in the 1980s. When events are less likely to occur, synfuel costs more likely to be high, or long-run oil prices are expected to be lower, the appropriate choice for the nation is a small or moderate effort. With such expectations, it becomes more likely that market incentives alone would not be adequate to motivate private firms to undertake initial coal-based synfuels projects. Although some production might take place, Table 6-14 results suggest that, without additional incentives or a program, the levels of production, learning, and infrastructure development might well be less than the nation may wish to have as a prudent insurance investment in improved capabilities with the synfuels option.

Using Sensitivity Cases

The sensitivity cases illustrate the use of the decision analysis model as a tool—not to determine one answer but rather to illustrate the implications of changing assumptions and relationships. The sensitivity cases illuminate which parameters, effects, and relationships are key determinants of relative program values and appropriate choices. Those critical specifications deserve special consideration and further study as part of deciding "How large?" an initial production effort is worthwhile.

NOTES

1. The distracting effects of inflation are suppressed because they are inappropriate for this cost-benefit assessment of alternative ways to allocate our limited resources. Since inflation affects all investment opportunities, the comparison among alternative investments should not be affected by performing the analysis in real terms.
2. For example, the base case assumes a probability distribution of 20 percent, 45 percent, and 35 percent for the basic synfuel cost outcomes of $35, $50, and $65 per barrel. However, if one believed the $35 cost outcome had only a 10 percent probability and the $50 and $65 outcomes occurred with equal probability of 45 percent each, then the relative values of Programs II, III, and IV would be 22.3, 30.3, and 22 respectively (billions of 1980$ net present value relative to no program reference value). The calculations draw from Table 6-2 and the stated probabilities. For Program III, the calculation is $[(.1) (81.9) + (.45) (45.0) + (.45) (4.1) = 30.3]$. With such example probability assumptions, Program III yields the highest expected value and is the best or optimal choice. Similar calculations that allow assessment of relative program values) are possible for any particular combination of outcomes and probabilities for synfuel costs or long-run oil price trajectories.
3. From Table 6-2, $10.2 billion positive net benefits for Program II exceed the negative net benefits for Program IV (−$33.6 billion) by about $44 billion.
4. Note that these are maximum expected costs only in a prospective, expected-value sense, assuming commitments must be made now. If future decisions can be flexible and respond to changing information, the maximum expected costs in the future should utilize whatever updated probability distributions are operative at the time commitments have been made. Assuming information has updated synfuel cost probabilities to be 100 percent for the high outcome, the expected cost of not modifying the program at that future point is the full difference between the expected value of Program II (+10.2) and Program IV (−33.6) or $44 billion.
5. If it could be obtained, the perfect information on long-run oil price trajectories has substantially less impact on optimal decisions and, accordingly, lower expected value. From Table 6-3, the expected value of perfect information about long-run oil price trajectories is about $5 billion NPV (0.25 × $[21.1 − (−.2)] = 5.3$).

6. For example, results for particular combinations of outcomes for synfuel cost or oil price trajectories will not be displayed or discussed. For more detailed model results for the various sensitivity cases see Harlan (1981), especially Appendix IV-C.
7. Note that program benefits for Case D and Case E are essentially identical. Case D, a 30 percent probability for an *SDE-S* event with zero probability for a price jump event, is essentially equivalent to a 10 percent probability for *SDE-S* combined with a 15 percent probability for *SDE-P* events.
8. However, two observations bear notice. First, the stylized price jump events are fairly long-term developments where price jumps do not decay rapidly. They are not temporary supply interruptions. Second, the price trajectories realized during frequently repeated *SDE-P* events are extraordinarily aggressive and would likely cause severe economic damage to both industrialized and developing economies as well as enormous international transfers of wealth. It is not implausible that repeated economic damage and wealth transfer of the magnitude implied by repeated or high probabilities for price jump events, particularly if they were the result of conscious political manipulation, would be destabilizing to the international political order. The price trajectory embodied by repeated *SDE-P* events is— in real, constant 1980$ per barrel—$72/bbl (1990), $85/bbl (2000), and $94/bbl (2010) for the medium long-run oil price trend. High probabilities for price jump events pose a very dire future for the world and should not be ascribed without careful consideration of the many implications for worldwide economic growth rates and the stability of the international order.
9. Arguably, there is some positive correlation between the existence of deployability constraints and levels of factor cost inflation. Factor cost inflation arises because of deployment constraints. Accordingly, one could argue that factor cost inflation would be reduced in a case such as Case G, where future deployment capability is independent of previous levels of production. The results of a sensitivity run where factor cost inflation occurs in the 1980s only are +12, +10, and +1 for Programs II, III, and IV, respectively. With factor cost inflation absent in all period, the results are +36, +48, and +67.
10. One might expect Program II to be the best program if deployability effects were completely eliminated. However, two effects seem to maintain a slight advantage for Program III relative to Program II. First, the expected value of learning is about 9 percent greater for Program III compared to Program II (.0795 percent as compared 0.073 percent). The greater learning may well offset the added costs from factor cost inflation (10 percent for Program III). Second, the direct production benefits are still maintained for Program III, despite a modest amount of added costs from factor cost inflation. If, however, larger programs provide the same amount of learning as smaller programs *and* deployability effects are absent, Program II yields

benefits quite similar to those of Program III. Model results for such a sensitivity experiment are +39.4, +40.5, and +24.3 for Programs II, III, and IV, respectively. Thus, if learning is the primary concern, a "small" program could be as good as the "moderate" program.

11. Figure 6-7 also illustrates the relative insensitivity of the general trend of results to the assumption for the BDC parameter. The base case assumption for BDC of twenty nominal plants or 1.0 MMBD per decade, yields results that are reflective of results for either relatively high (BDC = 40) or low (BDC = 10) assumptions for base deployment capability. The effect of the base deployment capability assumption is seen most strongly in the absolute benefits associated with the "no program" reference case. Table 6-8 shows that the "no program" zero point occurs at 19.3 if BDC equals ten projects (0.5 MMBD perdecade) or at 66.9 if BDC equals 40 (2.0 MMBD per decade).

12. Specifically, Programs II and IV are assumed to have the same benefit outcomes as the base case values assumed for program level III. Note that even with the same expectations for the DIF outcome for all programs, the linear modeling specified for expanding deployment capability causes future maximum deployment capabilities to increase in direct proportion to the size of initial production programs.

13. However, factor cost inflation alone is not responsible for the relatively small or negative additional benefits observed in many sensitivity cases for Program IV as compared to Program III. Even without factor cost inflation, the results exhibit declining marginal returns to program size. Marginal benefits per plant are 5.8 (from Program I to II), 5.1 (II to III), and 4.4 (III to IV).

14. As might be expected, limitation of factor cost inflation to one decade only (Case Q) increases slightly the expected value of the additional deployment capabilities purchased via larger initial production efforts. Program IV is 12 percent better than Program III in Case Q, while only 2 percent better in the base case.

15. This estimate is derived by straightforward lifecycle accumulating with discounting. A nominal plant produces about 50,000 barrels per day or 18.2 million barrels per year. Additional costs of $5 per barrel increase annual costs by $91 million. Over thirty years with discounting at 5 percent, this totals (via the discounting summary factor of 15.4 for a constant stream of values over 30 years at 5 percent) to $1.4 billion present value in the year of initial plant operation. For plants beginning operation in 10 years (1990), the $1.42 billion translates to a present value in 1980 of $0.86 billion.

16. Since different synfuel cost outcomes are possible, the expected (probability-weighted) additional costs from fractor cost inflation depend on the percentage amount of factor cost inflation and the probability distribution for various synfuel cost outcomes. For the base case specifications,

THE DECISION ANALYSIS RESULTS   221

the probability-weighted or expected average synfuel cost outcome is $52.25 per barrel. The expected synfuel cost level ($52.25/bbl) is 4.5 percent greater than the $50 level in the example. Thus, the per plant additional costs would also increase by about 4.5 percent from $0.86 to $0.90 billion.

17. The calculation is: $(99.3 - 29.2) = 70.1; 70.1/18.0 \times 10\% = 39\%$.

18. The estimates in Table 6-11 slightly overestimate the crossover levels of factor cost inflation required to make equal the net benefits of alternative programs. This mis-estimation occurs because the values calculated in Table 6-11 reflect factor cost inflation only for plants built in the 1980s. The effects of factor cost inflation on plants built in later decades (1900 or 2000-2010) are not reflected. Sensitivity Case Q shows that the effect of limiting factor cost inflation to one deployment period only is to increase the relative value of larger initial production efforts. The difference between the benefits of various programs is increased leading to higher estimates of crossover factor cost inflation. However, sensitivity Case Q also suggests that the runs with factor cost inflation in one period only provide a fairly accurate indication of relative program values. Thus, values in Table 6-11, while not exact estimates of the crossover amounts of factor cost inflation that would equilibrate programs if factor cost inflation occurs in all deployment time periods, provide a good and somewhat overestimated (and thus conservative) indication of the crossover levels.

19. *An Example Use of the Plot in Figure 6-12:* Suppose one had a 25 percent probability for price jump events and 10 percent probability of security events. Suppose also that Program IV is expected to have 25 percent factor cost inflation (FCI) and Program III 10 percent. From Curve C, the crossover level of FCI for Program IV compared to Program II is about 22 percent; thus, Program II is better than Program IV. From Curve B, the crossover FCI for Program III compared to Program II is about 16 percent. However, since Program III is expected to incur only 10 percent FCI, it is better than Program II. Program III, therefore, is better than both Programs II and IV and is the best choice. The intersection of the verticle line at 25 percent probability for *SDE-P* events and Curve A defines the amount of FCI for Program IV that would cause Program IV to become worse than Program III (assuming no FCI for Program III). For this example, that intersection occurs at about 12 percent FCI for Program IV.

20. Comparison of Case O/E with Case K/E shows that if basic existing deployment capability is relatively small (BDC equals ten in Case K/E compared to twenty in Case O/E), the crossover levels of factor cost inflation increase about five percentage points for a large program compared to a moderate program. Factor cost inflation is likely to increase strongly if base deployment capability is relatively small. Thus, a moderate program (III) may be worth the factor cost inflation that may be incurred, but the added factor cost inflation for a large program may more than offset any added benefits.

21. There is an additional qualitative difference in decision analysis results when the length of decline period (LDP) is less than ten years. When LDP is greater than the time required to deploy synthetic fuels (plus the time after event occurs but before a decision is made), decisions on production expansion subsequent to an event will reflect the fact that the occurrence of an event increases the expected value of plants decided upon and built after the event. If the LDP is short, an *SDE-P* event is over and oil prices have returned to their long-run trend before plants built as a result of decisions after the event can come on-line to produce synthetic fuels. When LDP is short, deployment decisions are not made in response to events but, rather, because of the prospective expectation that an event may occur and increase the expected cost of oil and value of synfuel production. When LDP is long enough to extend beyond the date at which plants built in response to the event come on line, deployment decisions reflect both the occurrence of an event and expectations for future events.
22. If Case Z specifications are run with no factor cost inflation, results illustrate a situation where net program benefits exhibit increasing and constant returns to scale. Benefits from Programs II, III, and IV in such a no factor cost inflation, high SPL case are +31.5, +67.4, and +135.2, respectively. The marginal benefits per nominal plant in going from program levels I to II, II to III, and III to IV are 6.3, 7.2, and 6.8, respectively. Thus, marginal benefits show increasing returns to scale between program levels II and III and nearly constant returns to scale between Programs III and IV. These results suggest that much of the observed effect of declining marginal returns to increasing program scale observed in the base case can be explained by the combined effect of factor cost inflation and synfuel production limits. However, in a smooth world with no surge deployment events, no factor cost inflation, and high (8 MMBD) synfuel production limits, net program benefits exhibit declining marginal returns to program scale. Marginal benefits are 3.2 (program levels I to II), 1.7 (II to III), and 1.4 (III to IV).
23. Individual private firms also may have different attitudes (subjective evaluations) toward gains or losses. For example, large losses may be weighted more heavily than large gains. This phenomena—risk aversion—may yield different risk weighted decisions than those that might be made by a risk neutral decision maker who placed equivalent subjective weights on large gains or large losses. Risk aversion, while recognized as a possible effect, will not be the subject of further sensitivity assessment or discussion in this study.
24. The calculation runs as follows: 5 percent rate time preference for consumption by society and corporate stockholders translates via a 15 percent tax rate for individual stockholders earning to a post-tax corporate return of 5.88 percent (5/0.85 = 5.88). If the corporate tax rate is 45 percent, the

corresponding pre-tax interest rate is 10.7 percent $(5.88/(1 - .45) = 10.7)$. The marginal tax rates are after CONAES Modeling Resource Group (1978: 14). For a good discussion of differences between discount rates applied to social decisions and private investment decisions, see CONAES Modeling Resource Group (1978: 14-15).

25. Actually, sensitivity case specifications cause Case *EE* results to reflect the value of information for later deployment decisions in addition to the direct value of energy production. Thus, Case *EE* results overstate the expected private choice, even if only the simple accounting value of the energy produced were considered by private firms.

26. The absolute amount of benefits for Cases *DD*, *EE*, and *FF* have little significance, because the modeling assumptions used to eliminate deployability effects allow large synfuel production expansion irrespective of prior production. Only the relative benefits across programs within a given case have significance.

7 SUMMARY OBSERVATIONS ON "HOW LARGE?" A SYNFUELS PROGRAM

Decision analysis results vary according to particular probability and parameter assumptions. However, the range of decision analysis results support three major summary observations on the "How large?" question.

BIGGER IS NOT NECESSARILY BETTER

In general, decision analysis results indicate that net benefits do not increase in direct proportion to program size. Often, larger programs, oriented toward production volume, are expected to yield fewer benefits than more modest efforts that provide basic information and learning as well as some infrastructure development. The base case illustrates declining marginal returns to program size. The "moderate" program (III), although twice as large, is only 40 percent better than the "small" program (II); the "large" program (IV) is only 2 percent better than the moderate program (III) half its size. For many other sensitivity cases, smaller programs are better than larger efforts. For example, in Case E, where oil price jump events occur with 15 percent likelihood, a small or moderate program yields essentially the same level of benefits and both are much better than a large program.

These diminishing marginal or absolute benefits with increasing program size are obtained despite analysis relationships that expect larger initial production programs to yield somewhat more learning and substantially more infrastructure development than smaller efforts. In scenarios where synfuels should be aggressively deployed, such improved capabilities could substantially increase the relative value of larger initial production efforts. Since information and learning benefits from initial synfuels production efforts do not increase strongly with increasing program size, however, the major reason and advantage for larger programs is to exercise the engineering-industrial infrastructure more extensively in order to build more deployment capability. (For some views of oil prices and synfuel costs, the additional volume of energy production also increases net benefits.) The added deployment capability from larger initial production efforts can be valuable in some scenarios, but the added near-term costs of factor cost inflation must be borne in all scenarios over a large volume of initial production.

On balance, it appears that the added costs and risks of a large program (IV) generally more than offset the added deployability (and learning) benefits. Bigger is not necessarily better. Similar relationships hold qualitatively for a moderate program (III) as compared to a small effort (II). Diminishing marginal benefits with increasing program size are observed. However, a moderate program provides added learning and some infrastructure development while experiencing less factor cost inflation than a large program over a smaller volume of (possibly) subsidized initial production. For a moderate program, the added benefits from improved surge deployment capabilities are more likely to exceed any added costs. If such insurance capabilities are reasonably likely to be utilized, the added costs and risks for a moderate program may be worthwhile compared to a small program that provides only basic information and an initial increment of learning.

PREFERRED PROGRAM SIZE IS SENSITIVE TO A FEW KEY ASSUMPTIONS AND EFFECTS

Decision analysis results should and do change with varying probability and parameter specifications. Relative program benefits and preferred choices are much more sensitive to some assumptions than to others. Those specifications having the greatest effect on appropriate answers to the "How large?" question merit special attention. In particular, four key assumptions and effects have a strong influence on relative program values: (1) synfuel production costs relative to long-run oil price trends; (2) the likelihood and description of surge deployment events; (3) factor cost inflation; and (4) the importance of infrastructure development.

Synfuel Costs Relative to Long-Run Oil Prices

Tables 6-2 and 6-3 illustrate the strong implications for appropriate decisions of varying outcomes for synfuel production costs and oil price trends. Synfuel costs and underlying long-run oil price trajectories have a critical effect on relative program values, even if oil price jump and security events are fairly likely to occur. If costs for coal-based synfuels are "low" ($35 per barrel) or oil prices are likely to follow a high long-run trend, no production subsidies are required, even for initial projects. Larger production levels are favored. However, if synfuel costs are certain to be "high" ($65 per barrel), the appropriate choice varies directly with oil price trends, ranging from "no program" if oil prices are expected to follow a low long-run path, to a moderate program if the oil price future is clearly a high long-run trend with price jumps above that high smooth trend also quite likely. Moderate outcomes for synfuel costs or oil price trends favor intermediate choices for program size.

Likelihood of Surge Deployment

Sensitivity cases for the probability and description of surge deployment events show that program preferences shift markedly with different assumptions. High probabilities for severe price jump (*SDE-P*) or security (*SDE-S*) events favor larger programs and increase the expected benefits from synfuels substantially. Low probabilities for price jump events reduce synfuel benefits for all programs, with a small program (II) frequently emerging as the best choice. On balance, a moderate program (III) is the best choice over a wide range of probabilities and descriptions for surge deployment events.

Factor Cost Inflation

Net benefits and relative program values are very sensitive to assumptions for factor cost inflation. If factor cost inflation is absent for all production levels (and coal-based synfuel costs are moderate), a large effort (IV) may be the best choice, even if surge deployment events are unlikely. However, Chapter 3 has shown that factor cost inflation is a plausible effect that could increase substantially costs for larger initial production efforts. Because factor cost inflation is likely to be greater for larger programs and is incurred over a larger volume of initial production, it reduces the expected benefits of larger programs much more than those of smaller programs. Factor cost inflation can cause small or moderate initial production efforts to be much more cost-effective than larger programs. For many assumptions about the likelihood of surge deployment events and other factors, the levels of factor cost inflation likely to be associated with a large initial production effort more than offset any added benefits. Factor cost inflation can make a large program worse than more modest efforts or no initial production program at all.

Value of Infrastructure Development

Finally, the value of infrastructure development bears importantly on the relative attractiveness of larger initial production efforts. The main advantage of larger programs is that they build infrastructure for expanded future synfuels deployment capability. The value of such enhanced deployability depends on how large the market role for coal-based synfuels production will be at the energy price levels necessary to make such production economically attractive. If domestic supply and demand responses or other constraints act to limit the maximum role for coal-based synthetics through 2010, the value of the infrastructure development purchased via larger initial production efforts is reduced. Conversely, if the role for coal-based synthetics is likely to remain large, the value of additional deployment capabilities obtained via larger initial production efforts significantly increases. Our expectations for the future role of various energy sources should affect our near-term decisions about how much deployment capability to "buy" as strategic insurance.

The value of infrastructure development is also affected by outcomes for the linkage between production experience and expanding future deployment capability. This linkage is not well understood. A strong linkage makes larger initial production efforts more attractive, since they can buy more deployment capability more quickly. However, the need to obtain infrastructure development via larger programs may be reduced if a low synfuel production limit is likely to make large deployment capabilities unnecessary or if levels of existing deployment capability are already considered ample.

MODERATE PROGRAM IS ROBUST CHOICE

Many sensitivity cases for key probabilities and parameter assumptions, as well as many descriptions of program cost and benefit relationships, show program level III as the optimal choice across a wide range of expectations. For example, Table 6-7 summarized sensitivity results showing that a moderate program (III) is optimal for a range of probabilities for standard oil price jump events from 15 to 50 percent. Expectations for shorter or less severe price jump events would extend the probability range over which Program III is preferred to include higher probabilities for price jump events.

Sensitivity results for factor cost inflation show that the crossover levels of factor cost inflation that would make Program III worse than a small program (II) are fairly substantial (30 percent for base case specifications). Moreover, if such levels of factor cost inflation were realized for a moderate program (III), Program IV—although twice the size of Program III—would have to experience only one-third more factor cost inflation before it would be worse than either a small program (II) or a moderate program (III). The relationships characterized in the sensitivity results for factor cost inflation (Table 6-11) suggest that Program III is likely to be a preferred choice over a wide range of both absolute amounts of factor cost inflation and also relative amounts for various program sizes.

Finally, sensitivity cases for the relative role of various sources of program benefits indicate that a moderate program (III) is a good choice. When either learning or deployability effects are examined in isolation from other benefits (see Table 6-8), Program III emerges as the highest valued choice.

A "moderate" program is not a particular production volume as much as a program that strikes a good balance between two key contrasting effects for the "How large?" question—factor cost inflation and infrastructure development. These effects are two sides of the same coin. Deployment constraints cause factor cost inflation for larger initial production levels. However, larger initial production programs more extensively build future deployment capability. As the probability of using maximum surge deployment capability increases, it becomes worthwhile to accept higher levels of factor cost inflation in order to purchase more insurance capability. A moderate program builds sufficient infrastructure without incurring unacceptably high levels of factor cost inflation. Technically, a moderate program (III) is defined as 500,000 barrels per day of coal-based synfuels production capacity by 1990. However, focusing on production volume alone can be misleading and counterproductive. Despite the technical definition of the moderate program, the sensitivity analyses suggest that the appropriate level for initial production insurance investments is exceeded once factor cost inflation adds near-term costs more than about 10 to 20 percent.

Overall, a moderate program appears to be a good or robust choice across a wide range of expectations. A good indication of the robustness of Program III may be developed by examining the full range of sensitivity cases presented earlier. One measure of the "foregone benefits" or losses of not selecting the best program choice is the percentage difference between the expected benefits of a particular program level and the benefits of the program with the highest benefits, according to the decision analysis results. For example, if Program IV had the greatest expected benefits with an expected value of 50, then, technically, it would be the optimal choice. If Program II had expected benefits of 40, the percentage benefits foregone from choosing Program II rather than the optimal choice (Program IV) would equal 20 percent [(50-40)/50 = 20%]. If Program III yielded expected benefits of 45, its percentage of benefits foregone would be 10 percent.

Table 7-1 summarizes such "foregone benefits" percentages for most of the sensitivity cases presented in Chapter 6. Program III is the best in fourteen of the twenty-nine sensitivity cases tabulated. Also, the average percentage benefits foregone figure for Program III is 19 percent. The tabulation shows that Programs II and IV are not nearly as robust as Program III. Program II is best in four of twenty-nine cases examined with an average foregone benefits percentage of 26 percent. Program IV is best in 11 of the twenty-nine sensitivity cases tabulated. However, Program IV has a much larger average foregone benefits percentage of 89 percent, due to large losses in cases where surge deployment capabilities are not likely to be utilized. In summary, program level III is more "robust" because it is more likely to be the correct choice and because the costs of being wrong are lower.

Table 7-1 illustrates another important point. Across the entire range of sensitivity cases examined, the "no program" choice is never the best choice. Thus, across the wide range of views of the world examined as sensitivity cases, at least a small initial production effort is better than no practical production experience at all during the 1980s.

Table 7-1. Summary of Sensitivity Cases results for "Benefits Foregone" Percentages (loss in benefits from choosing programs other than optimal choice from decision analysis model).

Sensitivity Case Label	Description[a]	Optimal Program Choice and Expected Benefits (Billions of 1980$ NPV)	Percentage Benefits Foregone from Choosing Program Other than Optimal (measured relative to value of optimal choice)		
			II	III	IV
A.	Standard base	IV: 45.2	30%	2%	—
B.	No events (SDE-P or SDE-S)	II: 10.6	—	76%	426%
C.	No SDE-P; 10% SDE-S	II: 14.1	—	35%	257%
D.	No SDE-P 30% SDE-S	III: 20.6	2%	—	105%
E.	15% SDE-P; 10% SDE-S	III: 20.2	2%	—	107%
F1.	30% SDE-P; 10% SDE-S	III: 30.5	18%	—	39%
F2.	25% SDE-P; 30% SDE-S	III: 38.3	24%	—	13%
G.	"Pure" learning effect	III: 41.1	9%	—	37%
H.	"Pure" deployability effect; no learning	III: 29.2	30%	—	7%
J.	"Pure" deployability effect; 8% learning	IV: 43.9	26%	1%	—
K.	BDC = 10	IV: 45.5	43%	11%	—
L.	BDC = 40	III: 47.5	21%	9%	—
M.	All programs with same benefit outcomes	III: 44.5	17%	—	9%
N.	Direct Production benefits only	III: 21.6	11%	—	63%

Table 7-1. (continued)

Sensitivity Case Label	Description a	Optimal Program Choice and Expected Benefits	Percentage Benefits Foregone from Choosing Program Other than Optimal		
		(Billions of 1980$ NPV)	(measured relative to value of optimal choice)		
			II	III	IV
O.	No factor cost inflation	IV: 99.3	68%	48%	—
P1.	Slightly higher factor cost inflation	III: 38.7	19%	—	11%
P2.	Double factor cost inflation	III: 29.6	6%	—	126%
Q.	Factor cost inflation in 1980s only	IV: 78.0	32%	11%	—
R.	50% oil price jump if SDE-P occurs	III: 26.5	13%	—	59%
S.	75% oil price jump	III: 35.2	23%	—	21%
T.	125% oil price jump	IV: 62.5	43%	16%	—
U.	Length of SDE-P decline period (LDP) 3 years	II: 17.7	—	7%	147%
V.	LDP = 10	III: 34.3	24%	—	22%
W.	LDP = 20	IV: 58.9	41%	14%	—
X.	More severe security event with 10% probability; no price jump	II: 18.5	—	2%	127%
Y.	More severe security event with 10% probability; price jump with 50% probability	IV: 62.6	44%	16%	—

Table 7-1. (continued)

Sensitivity Case Label	Description[a]	Optimal Program Choice and Expected Benefits	Percentage Benefits Foregone from Choosing Program Other than Optimal		
		(Billions of 1980$ NPV)	(measured relative to value of optimal choice)		
			II	III	IV
Z.	High synfuel production limit (SPL)	IV: 66.1	50%	23%	—
AA.	Low synfuel production limit (SPL)	III: 31.0	15%	—	30%
BB.	Stronger infrastructure development effect	IV: 64.5	44%	14%	—
Number of cases in which program best of 29 total cases			4	14	11
Average foregone benefits percentage			26%	19%	89%

[a]See tables and text of Chapter 6 for fuller description of various sensitivity case specifications.

CONCLUSIONS FOR THE "HOW LARGE?" QUESTION

Given our uncertain energy future and the insurance character of key benefits from initial synfuels production efforts, there can be no single answer to the "How large?" question. Answers and appropriate choices depend on poorly understood relationships and subjective personal expectations for the future. However, the decision analysis results provide some feel for expectations about key parameters, probabilities, and relationships that make various amounts of initial production seem worthwhile.

In general, at least a "small" program is better than no initial production experience, because it provides basic information about the energy price levels at which coal-based synfuels become economic. A small program also provides an initial increment of cost-reducing learning. Often, a "moderate" program improves upon a "small" effort by providing additional learning as well as stronger infrastructure development. If these additional surge deployment capabilities can be obtained with only modest levels of factor cost inflation, a moderate program is likely to be better than a small program for many views of our energy future.

In contrast, a "large" initial production effort often can be worse than more modest programs or no initial production of coal-based synfuels in the 1980s at all. A large program is a preferred choice primarily if synfuel production costs are low enough to make production subsidies unlikely, if factor cost inflation is not a serious effect, or if major oil price jump events or security events are very likely to occur. Larger programs provide greater increases in deployment capability. However, the enhanced deployability is worth the added costs from factor cost inflation only if major surge deployment events are thought to be quite likely.

The decision analysis assessment has addressed the "How large?" question for the nation as a whole. Assessing what private firms might do in response to market incentives alone and what size government program (if any) is appropriate requires detailed understanding of the incentives and decision factors of private firms. However, sensitivity cases examining plausible differences between benefits to the nation and incentives of private firms suggest that reliance on private marketplace decisions alone could result in less initial production experience than appears to be worthwhile for the nation as a whole. Special government incentives may be appropriate since coal-based synfuels may be in that narrow window where production costs are too high to allow the private sector to move ahead alone but low enough for the important energy option to play an economic role for some plausible, albeit unfortunate, developments in world oil markets.

Given the many uncertainties facing the synfuels investment decision, the attractiveness of the moderate answer to the "How large?" question is not surprising. It is prudent to move ahead on an insurance capability basis until near-term costs for production subsidies (the insurance premium) begin exceeding expected benefits. Ultimately, the appropriate amount of insurance investment depends on individual expectations for the future events that might require that surge deployment insurance capabilities be utilized.

III HOW TO DO BETTER?

Initial synfuel production efforts will require enormous investments of society's limited resources. Such investments should be carefully considered. Part I has addressed the questions of "Whether?" and "Why?" incentives and efforts beyond the free market may be warranted for synfuels. Part II has assessed "How large?" an investment may be appropriate. The answers developed in Parts I and II are not clear-cut. Net benefits, costs, and appropriate program size depend on several key uncertainties as well as on individual views regarding our energy future. Alternative perspectives on key parameters, probabilities, or values could transform one person's positive benefit-cost comparison into another person's negative assessment. However, Part I has shown that some initial production of coal-based synfuels is likely to be a good idea because of the insurance benefits of improved capabilities with the synfuels option should world events motivate a major surge deployment effort. Part II has characterized views of the world for which larger or smaller programs are appropriate.

Part III moves away from the quantitative emphasis of Parts I and II to address the qualitative issue of "How to do better?" with whatever size initial synfuel production program is undertaken. Cost-effective program design and implementation provisions can increase returns from initial production investments by emphasizing important sources of benefits or avoiding unnecessary costs. Even if net program benefits are negative or difficult to quantify, the value of initial production efforts can be increased by attention to important program purposes and benefits.

Part III develops a "cost-effective" approach to program design that has two major elements. First, the framework for program benefits, costs, and purposes is used to identify program design approaches that may accentuate program benefits or reduce expected costs. In particular, assessment of the "linkages" between initial production experience and the improved strategic capabilities resulting from that experience provides a basic framework that helps identify important program design issues. Second, since program design provisions that accentuate particular benefits may add costs or risks of costs, a cost-effective approach requires that the added costs of particular design emphasis or provisions be compared to the expected increase in program benefits.

This cost-effective approach raises many issues and suggests an extensive agenda for analysis to support efforts to increase the net benefits of whatever size initial production program is undertaken. For example, if learning benefits are to be emphasized, would two relatively small (25,000 bbl/day) facilities yield more learning than one nominal size (50,000 bbl/day) facility? The answer to this question is probably "yes," but a proper cost-benefit approach must further ask whether the incremental costs of smaller plants (which may not fully exercise potential economies of scale) are offset by an expected increase in learning benefits.

Attention to linkages between experience and improved strategic capabilities with the synfuels option also suggest issues for program design emphasis on strengthening infrastructure development for enhanced deployment capability. For example, if a key long lead-time deployment constraint is the number of architect, engineer, and construction (AE&C) firms capable of efficiently undertaking a multi-billion dollar synfuel project, should an initial production program preferentially involve some presently marginally qualified AE&C firms in order to "train" them? Sound decisions on this issue require assessments of whether AE&C services could be a limiting deployment constraint, of the means and extent by which initial production experience increases AE&C capacity, and of the likely cost increases due to the project design and execution inefficiencies that may result from using a project to train an AE&C firm that is not already wholly qualified to build a synfuels plant.

This approach for program design suggests a wide range of concerns. The range of potential benefits outlined in Chapter 2 suggests numerous opportunities and issues for increasing program benefits. Chapters 8 and 9 characterize program design issues for emphasis on two important sources of improved surge deployment capabilities—learning (Chapter 8) and enhanced deployability (Chapter 9). These strategic capabilities deserve special attention because, as sources of benefits relatively specific to synthetic fuels, they define important program purposes and can provide useful guidance for improving program design and implementation.

Part III does not attempt to present conclusive assessments of the many complex issues bearing on cost-effective program design. Rather, Chapters 8 and 9 raise broad questions and outline general analytical agendas for improving the design of initial synthetic fuels production efforts. Although many issues require further detailed assessment, some important design concepts are likely to remain robust in light of further analyses. These observations provide the basis for some concluding lessons of this study which are noted in Chapter 11.

The program design issues characterized in Part III assume that the purpose of an initial synthetic fuels production program goes beyond the simple production of energy to reduce oil imports. Part I has argued that for the purposes of reducing oil imports there is no special rationale for emphasizing synthetic fuels as compared to any other long-run alternative to imported oil. However, the insurance benefits of improved surge deployment capabilities provide a specific rationale for investments in major long lead-time energy options, such as synthetic fuels.

If the primary purpose of an initial synfuels production program is to meet some arbitrary production goal, then the program design concepts developed in Chapters 8 and 9 may not be appropriate. Clear definition of what we are trying to accomplish with initial synfuel production efforts has important implications for how we proceed to do better in moving toward the chosen objective. The preferred answer to the "Whether?" and "Why?" questions has strong implications for the "How to do better?" issue addressed in Part III.

As we will see in Chapter 10, the Energy Security Act of 1980 does not resolve possible tensions between a production volume purpose and a capability purpose for the major initial synthetic fuel production effort it authorizes. The tensions between program purposes must be addressed by the Synthetic Fuels Corporation as it develops its policies, program, and individual project selections. Fortunately, program purposes can be complementary to some extent. Attention to learning or deployability purposes need not preclude significant production volumes. However, production becomes a co-product or by-product of initial projects rather than the dominant objective and measure of the worth for a project or program. Part III shows how defining the purposes and objectives of an initial synfuel production effort as improving capabilities with the synfuels option suggests program design issues and approaches for doing better with initial synthetic fuel production efforts.

8 PROGRAM DESIGN ISSUES FOR ENHANCING LEARNING

INTRODUCTION: TWO TYPES OF LEARNING

Learning from initial synthetic fuels production experience should allow subsequent synthetic fuel deployment efforts to incorporate "better" or lower cost plants. Chapter 2 described two important types of learning—selection learning and technology improvement learning. Careful descriptions of both types of learning can suggest program design issues and project implementation approaches for increasing learning benefits.

Selection learning can provide information on the relative costs of different technologies, thus aiding decisions on *which* technology to use in a surge deployment effort. Poor technologies can be eliminated, thus reducing the average cost of the technology mix resulting from subsequent choices.

Practical experience with particular technologies should suggest technology design or project execution improvements that will lead to *better versions* of whatever technologies are utilized. This technology improvement learning should lower the relative costs for subsequent plants utilizing improved versions of technologies.

Both selection and technology improvement learning enhance understanding and capabilities with the coal-based synfuels option and should result in lower costs for surge deployment efforts. The distinction between these two types of learning is useful, however, because the linkages between initial production experience and improved surge deployment capabilities are different for each source of learning. Consequently, the program design issues based on these linkages have different emphases.

This chapter characterizes program design issues suggested by each type of learning. Issues are developed on a generic basis without detailed reference to any particular proposal for an initial synfuels production program. Selection and technology improvement learning are discussed separately but with a parallel organization. The basic process for learning and the linkages between initial production experience and improved information and capabilities are outlined first. These linkages suggest program design provisions for enhancing learning. Particular program design provisions raise many questions requiring further detailed analysis to determine whether such provisions are cost-effective.

SELECTION LEARNING: DESIGN ISSUES AND ANALYSIS

A variety of different technology and resource base combinations are now available to support commercial synthetic fuels production, and new technologies are under advanced development. Despite the availability of many options, however, there is limited practical commercial experience with many major alternatives. Consequently, limited information is available to assess which technologies should be selected for use if a surge deployment should become necessary.

Selection learning benefits stem from the value of information for improving future decisions. Experience from initial production efforts should clarify existing uncertainties about the relative cost and performance of different technology and resource base combinations. Such information may improve future selections, allowing emphasis on relatively lower cost technologies for later deployment efforts.

The process for selection learning suggests three major design principles. First, the information developed should facilitate *sound differentiation and choice* among major technology and resource base alternatives. Second, since selection information is to be applied to future deployment efforts, it should be developed primarily for technology and resource base combinations that are likely to remain *significant options* for future surge deployment efforts. Third, in order to usefully guide future deployment assessments, information from initial production experience should be developed and maintained in a form that is *available and interpretable* to future decisionmakers who may have to make assessments for surge deployment projects on short notice.

These three principles—sound differentiation and choice, significant options for the future, and available, interpretable information—provide a framework for assessing and improving the information outputs of an initial production effort. Being able to select better technologies for future deployment requires an accurate, relevant, and interpretable characterization of "What happened?" with initial production projects. Program design and implementation provisions related to each of these principles for improved selection learning are described below.

Sound Differentiation and Choice

Several design and analysis issues for selection learning can be illustrated by reference to a stylized hypothetical example. Suppose each of three technologies for coal-based synfuels (A, B, C) has a range for expected production costs. (If we associate different likelihoods for the different values of this range, we have described a probability distribution for production cost.) Because of our limited experience with synthetic fuels, it is quite likely that these cost ranges are rather broad and overlap substantially, or perhaps even completely. If the cost distributions overlap exactly, there is no reason to emphasize or de-emphasize any particular technology in the mix of projects comprising a surge deployment effort. However, after practical experience is obtained, the cost distributions for various technologies may be narrower as well as shifted relative to each other. Suppose practical production experience has shown that technology C is relatively high cost. Responding to this information, future deployment decisions should de-emphasize technology C. The average cost of the technology mix employed for deployment should be reduced even if no technology improvement learning has occurred for technologies A or B.

Figure 8-1 illustrates this stylized example of information which helps differentiate technology cost distributions for the selection learning process. Although the example simplifies a complex situation, design issues in two major areas are suggested—the timing for selection learning and the type of information required to realize selection learning.

Timing for Good Selection Information. Selection learning occurs when information becomes available that shifts cost distributions relative to each other, thus clarifying which technologies are more likely to have lower costs. Such information could become apparent at any point in the progress of an initial production project. For example, detailed design for a project may reveal that a technology costs much more than previously thought.[1] However, many uncertainties about technology cost and performance are clarified only via practical operating experience. Some information will be obtained over the course of project design and construction, but the greatest increment in selection information occurs after production operations begin. The timing for selection learning, therefore, is strongly linked to the timeframe for initial operating experience.

Synfuel plants take five to seven years to design and construct. Consequently, there is a substantial "hiatus" period before greatly improved selection information will be available to assist deployment choices among technology alternatives. During this hiatus period, deployment selections must utilize whatever interim selection information has been produced by initial production projects as far as they have progressed. The interim information should improve as projects progress but is not nearly as good as the information available some time after initial operations have begun.

The hiatus period for selection information suggests several issues. First, the sooner production projects can reach initial production, the sooner improved information can be available to support surge deployment choices. Secondly, since surge deployment could become necessary before projects begun in the early 1980s become operational, we should seek ways of improving the quality and availability of interim selection information. Improved understanding of the value of interim selection information should clarify the merits of alternative strategies, such as "select and commit now" as compared to deferring choices for a second group of projects until better selection information is available.

Figure 8-1(a). Hypothetical Cost Ranges/Distributions Before Experience

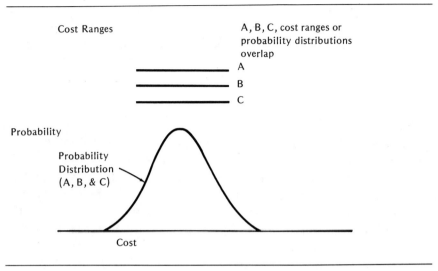

Figure 8-1(b). Hypothetical Cost Ranges/Distributions After Initial Production Experience

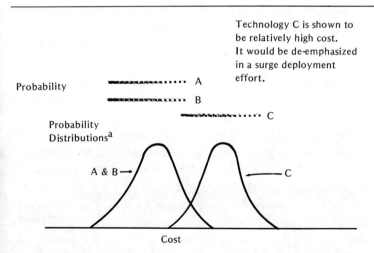

[a]Information from practical experience may also reduce the width (or variance) of cost ranges (distributions).

Thirdly, the hiatus period helps define practical timing for decisions on strategies and project selection for phased production efforts. Planning to realize selection learning should assess the coordination possibilities between decisions on second phase production projects and the quality of the selection information becoming available over the course of first phase projects. We should prepare ourselves to be able to choose well at any point, since a surge deployment effort could be forced upon the nation at any time.

Quality of Information for Selection Learning. The amount of selection searning is also affected by the number of projects and type of information obtained. Sound differentiation and choice requires quality information on the items necessary to accurately characterize and compare the performance of alternative technologies. If information from initial production experience does not foster well-grounded discrimination among technology options, the technology mix after initial production experience could be identical to the mix expected before experience. Selection learning benefits could be negligible.

Sound differentiation among technologies may not be possible when the technologies have similar costs and performance. However, poor structuring and management of the information available from initial production experience could obscure otherwise potentially useful information. For example, if the cost expectations for technologies A and B in Figures 8-1(a) and 8-1(b) are the same, experience from the initial production program should not suggest deployment emphasis (based on cost expectations) for technology A over technology B or vice versa. However, the program should develop information showing that technology C is relatively high cost, thus encouraging de-emphasis of that technology.

Several issues are raised by the stylized example illustrated in Figure 8-1. First, how many different projects with the various technologies A, B, and C are needed to yield good selection information? Suppose one project with each synthetic fuel process is undertaken. Will the information from comparisons of individual projects with each technology provide sufficient basis to confidently assert that technology C is relatively high-cost and therefore should be de-emphasized for future deployment? Or are two, three, or four observations with each technology necessary to provide enough discriminating cost information to provide a well-grounded basis to change the technology mix?

Related questions arise. For example, suppose we are reasonably sure that technologies A and B have the same fundamental cost distribution, due to underlying engineering similarities. If so, should we treat the technology as a generic technology, A/B, and avoid multiple, separate commercial projects with each technology individually?[2] Recall the stylized example of Figure 8-1. If the total number of projects is limited, would efforts emphasizing selection learning be more productively directed at discriminating between technology C and generic technology A/B rather than among all three technologies?

The quality of selection information also will be affected by the ability to distinguish between cost outcomes inherent to a particular technology and outcomes primarily related to the execution or management of a particular project. Just as a poor production of a play could inappropriately bias the audience's view of the playwright's script, poor project management could make a technology look more expensive than it would be in the hands of a different and better project management team. For example, a difficult site or poor project execution performance could result in a *project* utilizing technology B having higher costs than a project utilizing synthetic fuel technology A. This result could occur even though the underlying cost distributions for the two *technologies* were identical. Bad selection information could be counter-productive, actually increasing the average cost of the resulting technology mix.

For industry growth reflecting gradual market evolution, such project-specific outcomes may offset each other. A bad outcome with technology A in one project may offset a good result with the same technology in another project. However, selection information for surge deployment scenarios may not be able to reflect the advantages of a significant number of contrasting projects that can help distinguish technology-generic outcomes from project-specific outcomes. Given a limited number of projects, good selection information requires provisions to obtain information that will help distinguish between cost and performance outcomes related to the *technology* utilized and those related to *project execution*. Such information would include data on the component costs, labor rates, and construction schedule associated with the actual construction costs for a given project. This information would help characterize project-specific costs and outcomes and facilitate cross-project comparisons.

The need to separate technology performance outcomes from project execution outcomes further compounds the issue concerning the number of initial production projects needed for good selection information. More projects may be needed to provide a basis for sound differentiation and choice among various technologies.

Significant Future Options

Selection information may be of limited use, unless it is available for those technologies and resources that are likely to play a significant role in future surge deployment efforts. This principle suggests several design concepts.

Technology Diversity. An initial production program should be diverse enough to include major options that can be readily replicated for future widespread deployment. The number of significant future options helps define the number of projects necessary for adequate coverage. Diversity could be important, since failure to obtain experience with significant options could lead to a long-term technological bias against some alternatives. The development of options with cost or performance advantages might be delayed or completely stymied.[3] For example, if technology B in Figure 8-1 were excluded from an initial production effort, future deployment assessments might view it as less of a sure bet. The surge deployment technology mix might de-emphasize technology B, even though its fundamental cost distribution and expected costs were the same as technology A. Alternatively, if an initial production effort were not diverse enough to include major technology option C, then information that technology C was more costly might not be developed prior to a surge deployment effort. Technology C would be included in the surge deployment technology mix, thus increasing the average cost level.

Diversity is an important concern. However, it is not an end in itself. Diversity is necessary and valuable primarily for improving the basis to choose among significant options for future surge deployment. Given the limited number of initial production projects possible, overemphasis on diversity could reduce selection learning. For example, if one existing technology is very likely to be more costly than another existing technology, the higher cost technology may be so unlikely a candidate for future replication that it does not merit being included in an initial production program which should be oriented toward improving capability for the future. If we are sure now that a technology has little future role, it does little good, for the sake of diversity, to obtain better information on its unlikely role.

Diversity is primarily important for significantly different technology options. The definition of different options should be carefully drawn. Two different technologies may not necessarily constitute a significantly different option. For example, engineering similarities are such that Lurgi gasification followed by *methane* production may not be a significantly different option than Lurgi gasification followed by *methanol* production. Diversity fosters valuable information by clarifying choices among technologies that could have significantly different costs. If two technologies are so similar that they are very likely have identical costs, they could be viewed as one technology option from the standpoint of structuring an appropriately diverse initial production effort.

Resource Base Diversity. The principle of acquiring selection information for significant future options also raises the question of resource base diversity. A surge deployment effort requires that many projects be built simultaneously. Multiple projects would use many different coal resource bases. Unfortunately, different synthetic fuel processes often perform quite differently with different coals. In a sense, there may be a significantly different cost distribution for each different combination of a particular technology and a given coal resource base.

This fact complicates the technology selection guidance that may be available from an initial production program. For coal M, technology B may perform much better than technology A, while for coal N, the converse may hold. It is impractical to build a project representing each technology/coal resource combination, but information helping to assess the performance of different coals in different synthetic fuel processes can be developed. Laboratory correlations and experiments are possible. Also, for commercial production assessments, important information comparing different coal feedstocks can be obtained by shipping coals to a particular process plant and running the coal on a test basis. This "ship and test" approach has been used by several projects that tested the performance of particular coals at the commercial-scale Lurgi gasifiers of the Sasol facilities in South Africa.

The information from initial production projects may be more generally useful if provisions are made for obtaining information that will facilitate technology comparisons across different coals. The practical approaches for obtaining the relevant information require further assessment. It could be useful to structure initial production projects so that testing of alternative coals can be readily accomplished with minimal cost or interference to primary commercial production activities. Tensions and tradeoffs could arise with the commercial, production-oriented projects. However, program design should address the tradeoffs by assessing whether the incremental benefits of information on various coals are worth the extra costs and complications required to develop the information.

New Technology. Selection learning becomes less relevant to future surge deployment efforts as the technology for which information is available is replaced by new technology. In order to remain relevant to significant *future* options, selection information may have to be acquired for any new technologies likely to play a significant role in future surge deployment efforts. The research, development, and demonstration process may regularly produce new technology alternatives that, although ready for commercial-scale application, have uncertain costs and performance relative to existing technology. Depending on oil market developments and synfuel costs, widespread development of a growing commercial synthetic fuels industry may or may not occur. It is possible that a new technology, which was not included in an initial production program, could become available for commercial application before a synthetic fuels industry was generally economic and self-sustaining. In such a situation, selection information to support possible surge deployment could be improved by commercial experience with important new technology alternatives.

For example, it is unlikely that the so-called direct coal liquefaction technologies—such as solvent refined coal (SRC) or Exxon Donor Solvent (EDS)—will be ready for commercial use in the early 1980s. However, information about the cost and performance of the major technology alternative represented by direct liquefaction could be useful if surge deployment technology selection decisions became necessary in the 1990s and beyond. Such practical production experience may be primarily important if the new technology options are very likely to have lower costs than existing proven technologies.

Given the uncertain timing for possible surge deployment efforts, a synfuels production effort oriented toward maintaining insurance capabilities might be phased and extended to include projects incorporating economically attractive, but still-developing, technologies. Unlike a program concentrated into a single timeframe, a continuing capability program could assure that selection information remained "current" and relevant to significant future options. However, if synfuel costs and the oil market situation led either to an economic self-sustaining industry or made use of the synfuels option unlikely, the value of any continuing investments (subsidies) to keep selection information up-to-date should be critically questioned.

The dynamic element introduced by technological progress complicates selection learning and could cause the quality of selection information to decline over time. In terms of Figure 8-1(b), technological advance could yield several new synthetic fuel processes (E, F, G) with significantly lower cost distributions than technologies A and B. If this occurred, the selection information among technologies A and B would have little value, because it no longer would be relevant for improving choices among those technology options likely to be applied in future surge deployment efforts.

Available and Interpretable Information

The process for selection learning is based on improving future decisions via information transfers. For selection learning to occur, information must be available and interpretable to the relevant decision processes that will determine which technologies should be used in a surge deployment effort. Three major elements are important in achieving this objective. First, information that can help characterize initial production experience should be identified. Second, relevant information should be gathered and made available to assist future assessments. Finally, information should be maintained and transferred in a readily interpretable form.

Identifying Useful Information. Information that can enhance selection learning may not be identified or noted automatically as part of a normal commercial project. Engineers, plant managers, and companies struggling to build and run a synfuels project may not have the need or the inclination to recognize and record important items of information from an initial synfuels project. Much of the information is especially useful when compared with information from other projects. Individual firms may have little incentive to characterize their experience for the benefit of others. Such characterizations and comparisons, however, are fundamental to selection learning.

Firms often do not collect information important to characterizing their experience with a synfuels project. For example, several firms wishing to participate in a recent empirical study of cost estimation and technology performance for innovative chemical process plants were unable to participate because, in the course of normal commercial practice, they had not kept records on even the basic information required by the study.[4] Information important for selection learning may not be identified and noted unless advance provisions are made for obtaining such information as an important output of initial production projects.

Identifying the pertinent information is a straightforward process. We need only project ourselves into the future and ask what information from the 1980s experience must be known in order to characterize "What happened?" for various projects and what technologies are likely to have the lowest costs for future surge deployment efforts.

Making Information Available. Even if information useful for selection learning is gathered, its availability outside a particular firm is far from automatic. Information exchange about commercial production costs and other aspects of plant and project performance is not generally part of commercial firms' modes of operation. For example, although information on plant performance and throughput is essential to cost comparisons, such information is commonly closely held as proprietary data.

If the output of an initial production project is defined primarily as barrels of synfuel production and if information on project performance is held very closely, an initial production project may not make available the type of information that can guide and improve future technology comparisons. However, allowing initial production projects to proceed with normal proprietary information provisions may have favorable incentive value, since a firm that is more likely to profit from selling proprietary know-how from its initial project could have stronger incentives for efficient project execution and technology innovation. Yet, information important to project comparisons need not necessarily transgress on important proprietary values or positions. For example, information on total plant capital costs and construction labor rates, while important to cross-project comparisons, may have little proprietary value to the company undertaking an initial synthetic fuels project. Program design and implementation should address these tensions and tradeoffs for enhancing selection learning.

Maintaining Interpretable Information. Even if information for selection learning is identified, gathered, and made available, it yields benefits only if it improves the decisions of future decisionmakers at some uncertain future time. Unfortunately, we don't know either who those decisionmakers will be, or whether surge deployment project selections will be necessary in 1986, 1994, 2005, or ever. We can increase the likelihood that selection information will be useful by taking care to structure and maintain the information in a standard, well-documented, and readily interpretable format.

Specification of the exact form of information that characterizes initial production experience requires further careful assessment. However, provisions to assure that information from initial production experience remains available and interpretable are best initiated before or during initial production efforts. Otherwise, the potentially useful information from initial experience will be difficult to apply during the intense activity surrounding a surge deployment effort.

Analysis Steps for Improving Selection Learning

Cost-effective design and implementation of initial synthetic fuel production efforts should consider whether particular provisions to enhance selection learning are worthwhile. The process and principles for selection learning raise an extensive range of issues. Analysis to help illuminate "How to do better?" for selection learning includes three major elements, as summarized below.

Characterize Technology Cost Distributions. As illustrated by Figure 8-1, the fundamental process for selection learning involves distinguishing among cost distributions for important technology options. The first step in assessing and realizing the potential for selection learning requires characterizing which technologies are important and then describing the existing cost distributions for those technologies. The process of characterizing cost distributions should help identify which technology and resource base combinations are sufficiently different from each other to merit separate coverage in an appropriately diverse program portfolio.

Describe How Practical Information May Shift Cost Distributions. Selection learning occurs only if information from initial production experience is likely to help discriminate higher-cost from lower-cost technologies. Given the need to distinguish between technology-generic and project-specific cost outcomes, the difference between cost outcomes observed from initial production experience may have to be substantial in order to provide well-grounded differentiation among technologies. A major issue for selection learning concerns the number of projects and types of information necessary to provide a sound basis for comparing the relative cost and performance of significantly different major technology options.

These first two major analysis elements—characterization of cost distributions and descriptions of how they are likely to change—should provide guidance on the number and mix of projects required to obtain soundly based cost discrimination and selection learning. It is possible that the number of projects required could be large. In such a case, the potential for selection learning may be limited, and extensive efforts to realize unobtainable selection learning may not be worthwhile. On the other hand, discrimination among particular technology options could be both feasible and valuable. Program design provisions that emphasize selection learning only for those particular technology comparisons where differentiation is likely to be possible could strike a cost-effective middle ground. Further analysis is needed to help clarify opportunities, costs, and targets for enhancing selection learning.

Characterize Information Needed to Improve Future Decisions. Benefits from selection learning accrue when surge deployment technology selections are improved via information from initial projects. Selection learning can be enhanced if initial production efforts are structured to include information provisions to develop the types and forms of information useful to future decision makers. Analysis to develop these information provisions should include a detailed review of the technology assessment tools, project evaluation techniques, and information commonly applied for industrial decisionmaking. Effective methods for organizing, maintaining, and transferring information should be developed. In particular, information and techniques that help differentiate between cost outcomes due primarily to project execution from those that are technology-generic could be especially useful. Tradeoffs between proprietary interests in limiting access to project cost and performance information versus the value of freely available selection information should be explicitly assessed. In general, analysis should seek to develop a well-grounded information strategy that identifies useful types and forms of information and procedures for making such information available. Since information is an automatic by-product and key benefit of an initial production effort, such an information strategy could substantially increase expected benefits from selection learning while introducing only small additional costs.

Detailed Analysis Needed. Clearly, program design issues and provisions for enhancing selection learning define an extensive analytical agenda. Analysis based on the linkages between production experience and the ability to know better which technologies to use in the future can suggest program design provisions for realizing more selection learning. However, cost-effective program design requires additional assessments to compare the incremental value of a particular program provision to its incremental costs. Such tradeoff assessments should be based on understanding of the incentives, needs, and preferences of both the private firms whose pioneering projects will develop the information, and those who may need to use the information in the future.

Detailed assessment of this extensive analytical agenda for selection learning is beyond the scope of this study. Our primary emphasis is on the initial step of characterizing the basic principles, issues, and analytical approaches that suggest "How to do better?" for selection learning. A step-wise approach is necessary. The process begins with defining program purposes and benefits and proceeds by characterizing linkages between the production experience and the realization of benefits. It ends by detailed assessments and policy decisions regarding specific program design and implementation provisions.

TECHNOLOGY IMPROVEMENT LEARNING: DESCRIPTIONS AND DESIGN ISSUES

Synthetic fuel production requires orchestration of billions of dollars worth of engineering, material, labor, and other inputs. Practical experience with such a complex activity should help identify improvements that will allow subsequent efforts with the same or similar technology to be more efficient. Such "learning curve" effects lead to cost reductions that have been described in a range of qualitative and empirical studies,[5] and represent an important source of benefits from initial synfuel production projects. Examination of the process by which learning occurs and the linkages between production efforts and the generation and transfer of learning can suggest ways to increase the learning benefits from initial synfuel production efforts.

Program design and implementation provisions for enhancing learning should be based on a careful description of the learning process. Accordingly, this discussion begins with a general qualitative description of the process by which technology improvement learning occurs and is transferred to subsequent deployment efforts. This description suggests program design provisions or project attributes that could increase the learning outputs of an initial synfuels production effort.

An emphasis on learning benefits, however, could conflict with other program objectives, especially the attainment of substantial synfuel production volume. Clear definition of emphasis among program objectives is necessary to weigh any tradeoffs that arise. Also, some provisions for enhancing learning could increase the costs of initial production projects. Cost-effective program design requires that the incremental costs of particular program design provisions be compared to incremental benefits. This study notes major issues and related analyses that could clarify appropriate choices. However, a firm recommendation on whether a particular program design emphasis or provision is worthwhile requires more specific definition and analysis than is possible here.

The Process for Technology Improvement Learning

The process by which experience leads to technology improvements and cost reductions is a complicated and incompletely understood phenomena. The general outlines of the learning process, however, can be summarized from various written descriptions and from discussions with persons experienced with the "art" and process of technology design and improvement.[6]

Four Project Phases. Implementation of a complex synthetic fuel project takes place in several phases over several years. The learning process may be divided into four separate phases.[7]

The initial phase is project definition and plant design and engineering. This phase may last approximately one to three years. It is the period during which detailed project specifications are defined via site studies, preliminary and detailed engineering, and other paper studies.

The second phase is plant construction. This phase may overlap with project design to varying degrees and may last from two to four years or more, depending on plant size and complexity, the characteristics of a particular site, project conditions such as weather, labor availability and productivity, and so forth. Important skills during this phase include the project and construction management capabilities required to orchestrate the construction of a complex multi-billion dollar project. Specialized construction labor skills, such as high-pressure welding are also important.

Closely following completion of construction, the plant start-up and initial operation phase begins. During this critical phase, the integrity of plant construction and components is established, and initial production is attempted. Personnel responsible for plant design, construction, and operation become intensely involved in getting the plant started-up and operating to produce specified products. The length of start-up periods depends on plant complexity, degree of innovation, and the problems encountered. Start-up for a complex synthetic fuel facility may easily require six months to over one year.

Table 8-1 summarizes estimates for the approximate time required for implementation of these first three phases of a synthetic fuels project, excluding any preliminary planning, site acquisition, or permitting time. As the table shows, the physical and technological realities of designing and constructing a complex, multi-billion dollar, synthetic fuel facility make the initial operating experience unavailable for at least five to seven years. Front-end planning, acquisition, and permitting requirements could prolong this minimum timing by several years.

After start-up and initial operation, plant operations continue with emphasis on steady production and improving plant performance toward and beyond design specifications. During this early production period, plant operating personnel (as well as some design personnel) become familiar with operating techniques and the particular performance of the plant or technology. Such familiarity and experience may help plant operators identify ways to improve plant performance. Still, the dominant emphasis of the continuing production phase is on stable commercial operations to yield products. If all goes well, this fourth phase continues over the life of the plant, except for periodic shutdowns scheduled to perform necessary maintenance.

Differing Amounts of Learning in Each Phase. The likely amount of learning varies significantly during each phase of project execution. Sources of learning during the design phase include identification of relevant state-of-the-art design concepts, familiarization with regulatory and other performance requirements, interpretations of laboratory or pilot plant data, and so forth. Even without plant operation, some learning occurs via such familiarization with technology and because any given design can be improved by putting more time into it.

There is often substantial "art" in chemical engineering design for new technologies. Assumptions or "guesstimates" about the performance of an unfamiliar reaction, equipment component, or process configuration can be evaluated only by actual plant operation. Prior design (without operation) of a plant will allow subsequent designs to be somewhat better and save design time. However, most of the major process design improvements require practical operating experience to become apparent or proven.[8]

Table 8-1. Time Required to Implement a Commercial Synfuels Project.[a] (Years)

Phase	Nominal Schedule[c]	Accelerated Schedule[d]	Empirical Experience[e]
Design and engineering	2–3	1.5–2	3
Construction	3–4	3–4	2
Start-up and early operation	.5–2	.5–1	1
Approximate total[b]	5.5–9	5–7	6

[a] These estimates exclude times required for preliminary planning, site acquisition, and some permitting or other regulatory approvals.

[b] The total is only approximate since there is some potential for overlap among phases of a project. For example, some construction may overlap with the completion of detailed design. However, for first-of-a-kind plants, lack of experience makes extensive overlap of construction with detailed design more risky. See Merrow, Chapel, and Worthing (1979).

[c] Estimate by author developed by excluding permitting and approval times from generally available estimates for required project timing.

[d] Accelerated schedule reported in Fluor Corporation (1979), Figure 6-1 on p. 31. The Fluor accelerated schedule reflects substantial overlap of detailed design with construction. The estimate of one to two years for design reflects the amount of design time required before construction can begin. Fluor expects that a full three years are required to complete design. Consequently, there is little room for overlap between phases for the accelerated schedule reported by Fluor. Four and a half years is an optimistic minimum time to start up, excluding project definition as well as permitting. Start-up estimate supplied by the author.

[e] Estimates drawn from average actual time required for 23 innovative process plants as reported in Merrow, Phillips, and Myers (1981). Time for design and engineering includes one year for "project definition," which typically takes longer than scheduled for pioneering facilities.

Since the construction phase of a synfuels project requires orchestration of men, material, and equipment totalling billions of dollars, one would expect some learning to accrue as experience suggests improved construction techniques. The importance of such construction management improvements relative to process design improvements may well be limited, however, since construction management techniques for a large petroleum refinery or electric power plant are quite similar to those required for a synfuel project. Pouring concrete, welding pipes, or expediting materials are substantially similar whether the plant is to produce chemicals, electricity, or synthetic fuels.

The majority of learning with new chemical process plants occurs as part of the plant start-up and initial operation phase. During this third phase, one learns basically how well the overall technology and its various components work. Learning occurs in diverse, often unpredictable areas. Particular components of a synfuels process, which were artfully designed with limited information, may be revealed to be over-designed, under-designed, or mis-designed. Some parts of a production train may perform substantially better, worse, or differently than expected. The information can suggest modifications in either operation or equipment that can improve the performance of both the plant being started-up and subsequent plants using the same or similar technology. For example, information on the actual performance and capacity of process units should allow more precise design and equipment sizing. The savings on equipment investments should reduce per barrel capital costs. Also, "de-bottlenecking" will become possible as the capacity of process steps limiting production rates are identified. Other improvements in equipment design (such as special performance pumps) or process configuration could lead to improved performance and lower costs.

After start-up and initial operation, a commercial plant is generally operated with emphasis on stable production. If a plant operates smoothly without major changes in process conditions, learning accrues relatively slowly after start-up and the first year or so of operation. However, perturbations away from some base conditions often can reveal information about behavior and relationships within a system. Learning with a complex chemical processing facility is often greater when a plant operates near the edge or outside of its design regimes. (One reason that plant start-up can be a particularly informative period is that many parameters and process streams go through significant transients before a plant settles into steady-state operation at or near its design points.[9])

When a plant operates smoothly near its expected design conditions, not as much is learned about important process variables and relationships as when a process upset occurs.[10] Thus, when a plant works well, learning accrues relatively slowly during the fourth or continued production phase, unless a specific effort is made to experiment with process conditions. Some experiments may occur by accident. However, the level of experimentation with process conditions during the continuing production phase of a project will depend on the incentives and propensities of the particular firms that are the owner/operators of a plant. A firm's incentives for experimentation will also depend on the costs, risks, and expected benefits of using a particular facility to explore possible technology improvements as compared to the value of emphasizing maximum production volume.

Figure 8-2 summarizes these qualitative descriptions of the learning process as a stylized learning time line for an innovative chemical process project such as an initial synfuels plant. The horizontal scale indicates time during a project's evolution. The vertical axis reflects learning that has identified potential per unit cost reductions for a subsequent plant designed at a particular point through the progress of the initial project. The verticle scale purposely avoids indicating absolute magnitudes, since these are uncertain and vary with particular circumstances. The focus of the illustration is on locating the relative amounts of learning during different phases of a project's progress.

Figure 8-2. Illustration of Relative Amounts and Locations of Learning During Progress of a Synfuels Project

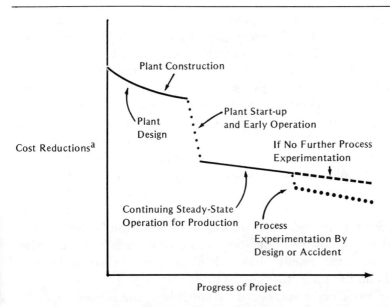

[a]"Cost Reductions" is a summary term for the ordinate variable. Actually the ordinate represents technology improvements identified at a particular point in the progress of a project which, if applied to a subsequent version of the same technology designed and built at that particular point in the progress of the initial project, would allow relative cost reductions of the magnitude depicted.

Figure 8-2 is distinctly different than the usual learning curve, which depicts costs as a function of cumulative production. Modest learning during the design and construction phases is reflected by the relatively flat slope. The start-up and operation phase is associated with a step-wise increase in learning. If a process operates smoothly without major changes in process conditions, learning accrues relatively slowly after start-up and the first year or so of operation. This is indicated by the relatively flat slope of the learning curve at the "continuing plant operation for production" phase. However, if a process upset or a change in process conditions occurs by accident or design (such as the plant owner/operator experimenting with different process conditions), then additional information and sometimes opportunities for design or operational improvements may become apparent. This is illustrated by the step-wise or discontinuous increment in learning.

Different Actors in the Learning Process. The learning process involves many different actors in addition to the personnel of the firm that owns and operates a synfuels plant. Implementation of a synfuels project involves several diverse organizations and their personnel. Plant design and engineering services are provided by "architect/engineering" (AE) firms. "Construction management" (CM) firms plan and manage the construction activity. Often the same firm providing the design and engineering services takes responsibility for construction management. "Component suppliers" (CS) provide discrete pieces of equipment or other commodities to a project. "Construction labor" (CL) services are obtained by construction managers from the general labor makret or particular labor organizations, such as unions (especially for skilled labor). Personnel from the "owning and operating company" (OOC) are involved in all phases. Their level of involvement is substantial in the project definition phase as well as in the plant start-up and operation phases. Often personnel for the owning/operating company are less involved in the construction management phases than most of the firms directly hired to build the project.

Each of these actors plays a different role in the process of technology improvement learning. For example, learning with generic components or unit operations of synthetic fuel facilities—such as coal preparation or sulfur removal units—accrues generally with the designers and operators of such units. Designers are located both in architect/engineering firms and firms supplying component equipment. Start-up and operating experience accrues primarily with the owning/operating company. However, relevant personnel from AE&C firms are usually integrally involved in the start-up and initial operation phase. Thus, the broad lessons of early operating experience are likely to become available to design personnel both inside and outside the owning/operating company. Design and construction management learning accrues primarily to the providers of those services—the architect, engineering, and construction (AE&C) firms.[11] Owner/operator purchasers of AE&C services may also learn how to better specify the services they purchase.

The different sources and actors for the learning process vary in their applicability and transferability from one technology or project to another. For example, design learning can be both generic- and technology-specific. Generic design learning might accrue when an improved coal gasifier design is developed that could be applied to a number of different indirect liquefaction technologies all requiring an initial gasification step followed by a liquefaction step to produce different products. In contrast, technology-specific learning might include design or operating improvements for a catalyst and reactor used only for a particular indirect liquefaction technology and product. Construction management experience is likely to suggest improvements that are generally applicable across a broad range of technologies and projects. Many aspects of the construction management process (procurement, scheduling, labor relations, etc.) do not change significantly for different technologies. In contrast, start-up and operating experience is relatively specific to a particular technology.

In order to yield cost reductions, the technology improvements identified via initial production experience must be transferred to future projects. The levels and transferability of learning depend on both the actors involved in an initial production project and the particular source of learning. For example, although design learning accrues with both the owning/operating company and the architect/engineering (AE) firm, the transfer of the learning to future projects may be predominantly accomplished by the AE firm. The project design learning accruing to the owning/operating firm may well be transferred to other projects that particular firm may undertake, but may not be as readily transferred to projects undertaken by other firms. Thus, the learning process for technology improvement involves both the *generation* or identification of technology improvements from various sources, as well as the *transfer* learning to future development efforts.

Table 8-2 summarizes descriptions of important attributes of the learning process for several different areas or phases of learning. Each different area involves different primary actors. Each has different actors that effectively transfer the learning to future deployment efforts. Each has different general applicability across technologies.

Key Observations from the Learning Process. The process by which technology improvement learning occurs is clearly complex, involving different sources, phases, and actors. However, several aspects of this general characterization for the technology improvement process have important implications for enhancing the learning benefits from initial synfuel production efforts.

First, the substantial increment in learning occurring only after initial operating experience has strong implications for appropriate timing and phasing. Second, learning occurs from basic experience with a project, which clarifies relationships and design parameters. Beyond whatever size is necessary to illustrate important technology performance relationships, the volume of product produced bears little relationship to the quality of the experience and information that provide the basis for identifying technology improvements. Thus, small commercial projects may be just as effective at generating learning opportunities as larger projects. Success at producing large production volumes does not necessarily generate learning that can improve technology for the future. Indeed, smaller projects may provide flexibility that fosters learning; problem projects could provide much learning but little synfuel production. Some issues and relationships, however, can be illuminated only by full-scale commercial projects. Both small and large commercial projects have advantages and disadvantages for learning, but production volume itself is clearly not a driving force for the learning process.

Third, many sources of synfuel technology improvements identified via one project are likely to be readily transferable to other technologies. This good potential for cross-technology learning

Table 8-2. Attributes of Learning Process for Synthetic Fuels

Area of Learning	Actors Involved in Learning Phase[a]	Actors Predominately Transferring Learning[b]	Degree of Applicability of Learning Across Different Technologies
General experience with synthetic fuels project	AE, CM, OOC (CS)[c]	AE, CM (CS)	Moderate
Technology specific design and operation	AE, OOC	AE	Low
Component design and performance	AE, CS (OOC)	AE, CS	High
Construction planning and management	CM, AE (OOC)	CM, AE	Very high
Plant start-up	OOC, AE (CS)	AE (CS)	Low
Operation	OOC (AE)	AE	Low

[a] Letter codes as follows: AE = architect/engineering firm, CM = construction management firm, OOC = owning/operating company for plant, CS = component supplier firms. (Construction management firms are often coincident with the AE firms.)

[b] This column indicates learning transfer to owning/operating companies that do not have previous synthetic fuel experience. Owning/operating companies may transfer substantial learning from all phases of their earlier projects to their own later projects.

[c] Parentheses indicate limited involvement in learning process or transfer to other projects.

expands the applicability and enhances the value of learning from initial synfuel production projects.

Finally, the incentives operating on various actors in the learning process can affect the amount of learning likely to be realized. Different actors have varying incentives to generate and transfer learning. Program design provisions that recognize and use varying incentives could enhance expectations for learning. The next section reviews some program design issues that are suggested by these four major observations.

Issues for Enhancing Technology Improvement Learning

Appropriate Phasing and Timing. Since technology improvement learning is fundamentally a sequential process, a phased program structure greatly enhances opportunities for learning. Moreover, the substantial learning occurring with start-up and initial operation suggests that the timing for program phases should be related to the time required for first-phase projects to provide good information based on sufficient practical operating experience. Major deployment efforts that are initiated before such learning has become available may repeat design flaws or forego design improvements. The hiatus period noted earlier for selection learning applies to technology improvement learning as well. Table 8-1 suggests that this hiatus period may continue for five to seven years after commitment to an initial production project.

The fundamentally sequential character of the learning process increases the value of proceeding expeditiously with the first few projects of an initial production effort. The sooner initial production experience is available, the better. Earlier experience from pioneering projects will reduce the hiatus period during which major technology improvements remain unidentified.

Moreover, earlier progress with initial production projects will provide earlier information on production costs and environmental issues. This information should clarify the value of continuing synfuels investments and indicate ways to mitigate any adverse effects of synfuels production.

An initial production effort could enhance learning by adopting a strategy of emphasizing a few "fast-track" plants in order to realize initial operating experience as quickly as possible. It could be worthwhile to accept the additional necessary costs in order to accelerate a few pioneering projects. For example, fast-track plants could require costlier expedited procurement and would incur risks from compressing design and construction phases. Also, fast-track plants might be expected to deliver exceptional environmental performance in return for expedited environmental approvals.

Although some fast-tracking may be possible and desirable for a subset of initial projects, efforts to accelerate project schedules for a large group of plants warrant further critical assessments. For example, while some overlapping of design and construction phases is common for well-known technologies, the risks of initiating construction before detailed design is substantially complete are much greater for first-of-a-kind facilities.[12] Attempts to overlap or short-circuit design, engineering, and construction phases could substantially increase costs and even prolong the time required to complete initial projects.[13]

The merits of accepting the added costs of fast-tracking depend on the extent to which accelerated learning from such projects could be applied to other projects proceeding at a more normal pace. If learning from fast-track start-up and operating experience can be useful to other plants in advanced design or construction, the value of a fast-track for at least a few projects is increased.[14] However, if information from expedited projects is unlikely to be available soon enough to substantially improve the normal-track projects of an initial production effort, then the added costs for special expedited projects may be less worthwhile.[15] Analysis to assess the merits of various fast-track provisions must be based on a characterization of the coordination possibilities between the time learning becomes available and the time it can be constructively applied to projects in various stages of development.

Smaller projects or a smaller first-phase initial production program might also reduce the hiatus period for first-phase learning. Design and construction of smaller projects could be somewhat quicker than that of larger plants. Also, a smaller initial production program would be less likely to encounter construction delays due to limited availability of key engineering and management services or equipment items. A larger program could strain the infrastructure for building synfuels plants in diverse ways ranging from poor labor productivity to bottlenecks for specialized equipment items. The resulting delays could prolong the hiatus period. In either case, learning could be delayed by emphasizing production volume from larger projects or more projects at the same time. Further analysis is needed to characterize these important tradeoffs between an initial production effort emphasizing production volume and one structured to enhance and accelerate learning.

Potential Advantages of Smaller Initial Production Projects. Large volumes of synfuel production are not necessary to obtain the basic experience that yields learning benefits. Many technology improvements can be identified via practial experience at a scale substantially smaller than a commercial project designed to fully exercise economies of scale. The ability to learn at relatively small scale is verified by the common practice in the chemical industry of starting small with initial commercial plants that incorporate innovative technology. A smaller plant is often built to test a technology and identify design or operational improvements that can be applied to later capacity expansions at the same or different sites. Many firms seek to "prove out" technologies at the smallest scale that will still provide the necessary information about technology performance for full-scale commercial production. Where learning is the major purpose for an initial production plant, production volume becomes primarily a means to an end.

Learning is a comparative process. Comparisons of performance between individual technologies or individual projects using the same or similar technology can be substantially more illuminating than experience without a comparative reference. Thus, two smaller commercial projects are likely to yield more learning than a single large project with the same or even larger production capacity.

The character of the learning process suggests that learning from initial production efforts could by enhanced by encouraging more smaller projects as compared to fewer larger projects. Several issues arise for cost-effective program design. Per barrel production costs and required subsidies may be greater for smaller projects than for full-size plants. Even if per barrel subsidy costs are greater for smaller projects, the total volume of subsidized production (and outlays for production subsidies) may be less. Thus, for a given amount of learning, smaller projects may require fewer subsidies than larger projects.

Other concerns limit and counterbalance the advantages of smaller initial production projects for enhancing learning. Some effects and relationships are illustrated only via practical experience with full-size projects. For example, without a few large projects, sound information about the economies of scale associated with synfuel production may not be as available. The production costs and environmental or socioeconomic effects of full-scale projects may not be adequately illuminated if only small projects are undertaken. Also, small projects could consume limited engineering and management resources without yielding as much synfuel production as possible.

Further assessment of the advantages and disadvantages of smaller initial production projects is an important issue for cost-effective program design. This limited review suggests that emphasis on enhanced learning can be fostered by favoring smaller initial production projects. However, the definition of "smaller," the strength of preference, and the value of some number of full-size projects requires further analysis and depends on preferences and tradeoffs among competing program benefits and purposes.

Cross-Technology Learning. Although technology improvements identified for a particular synfuel technology or process are most directly applicable to subsequent projects using the same basic technology, many of the basic building blocks and equipment items for various synfuel technologies are quite similar or even the same. For example, almost all indirect liquefaction or coal gasifications technologies include a processing step where sulfur-containing gases are removed. Improved sulfur removal technology would be applicable to many different technologies.

The many similarities among various synfuel technologies combine with the limited amount of practical experience available to increase the potential for cross-technology learning, especially in the 1980s. In general, the good potential for the transfer of learning among various technologies and projects makes it easier to realize learning and moderates the otherwise firm prescriptions for program structure and implementation provisions necessary to realize full learning benefits.

For example, because the existing experience base is so limited and the potential for cross-technology learning so good, learning during the design and construction phases of projects may be greater than it would be for a proven commercial process. Some AE&C personnel have noted useful "cross-talk" between design efforts proceeding simultaneously for several synfuel projects using different technologies.[16] When useful cross-talk between projects in different phases is possible, the advantages of having a tightly phased and sequential structure for an initial production effort become less compelling.

Good cross-technology learning also moderates the need to obtain experience with each of a diverse set of synfuel technologies, since experience with a limited set of projects may suggest technology improvements that are applicable to a wide range of synfuel technologies. For example, the coal-to-methane project beginning construction in North Dakota could provide basic experience that may clarify issues and identify technology improvements applicable to several other technologies for producing liquids from coal. The good potential for cross-technology learning may allow a smaller initial production effort, with necessarily limited technology and resource base diversity, to provide most of the technology improvement learning that could be expected from a larger program with broader coverage of major technology and resource base combinations.

Incentives to Encourage Learning. The many different actors involved in the learning process have different incentives for the identification of technology improvements and the transfer of learning to future deployment efforts. Moreover, different individual firms may have different propensities for generating and transferring learning. For example, an architect/engineering firm experienced with new technology design and development might identify more technology improvements than a firm that has built primarily standard plants using well-established technology. Design of an initial synfuels production effort should recognize the varying incentives and capabilities of the actors and firms involved in the learning process and perhaps utilize them to enhance prospects for learning.

Possibly the most important area for these differential incentives and propensities to advance learning may be for the firm (or group of firms) that owns and operates a pioneering synfuel plant. As illustrated by Figure 8-2, the amount of learning during the fourth or continued operation phase of a project can depend on the incentives of a plant operator. For a given commercial scale plant, a firm that owns a technology (and, hence, foresees some potential value from marketing the technology rights) may be generally more willing to experiment with process conditions than an owner/operator that has no economic interest in a technology other than to produce products from the particular facility. Put simply, a technology licensor is much more likely to experiment and learn with a technology than a technology licensee. The incentives of project participants to close the loop on technology evolution by developing and transferring learning to subsequent deployment efforts could aid program design and implementation to enhance learning benefits. For example, selection of project participants for initial production efforts might favor firms having the capability and incentives to aggressively seek technology improvements as well to transfer that learning to future deployment efforts.

Program incentives and project selection to encourage technology improvement must also address the issue of the level of "technological aggressiveness" that will be allowed or encouraged as part of initial commercial production efforts. A program oriented toward production rather than learning would incorporate on well-proven, low-risk technologies. Such a program would provide less technology improvement learning than a program that encourages some new technologies and accepts the attendant risks. A program emphasizing learning must strike a balance between innovation, risks of technological problems, and relative certainty of production volumes. The acceptable level of technical risk must be considered. Moreover, it is important to clearly articulate any learning objective at the outset, so that a project that produces little synfuels product but provides much learning will not be inappropriately viewed as a minor or "failed" project.

The level of emphasis on learning and the willingness to accept the risks of being technologically aggressive affect the types of projects and incentives that should be included in an initial production program. For example, relatively small commercial-scale facilities have great advantages for innovative or risky technology, since capital exposure is reduced if technical problems are encountered. Private industry often uses relatively small plants (sometimess less than commercial-scale) to reduce the capital costs and risks of experimenting with a technology.[18] Such small, innovation-oriented projects may require different incentives, risk sharing arrangements, and contractual structures. Further analysis is needed to characterize the many issues and tradeoffs that arise for considering whether and how an initial production effort should encourage innovative technologies.

"DOING BETTER" FOR LEARNING: FOUR LESSONS

Any initial production effort will provide some learning. However, a program designed and implemented to enhance learning is likely to be much different than a program that adopts production volume as its primary objective and measure of performance. Stylized choices between these two objectives are neither necessary nor appropriate. This chapter has shown that characterization of the learning process—including the linkages between how an initial production program is structured and implemented and how choices and technologies for the future are improved—can suggest program design provisions and issues for enhancing learning. Further analysis is needed to define cost-effective program design provisions. However, four major lessons for doing better with respect to both selection learning and technology improvement learning are summarized below.

A Properly Phased Program Greatly Enhances Learning

The technological realities of the learning process for both selection learning and technology improvement learning determine the time when most learning first becomes available. Learning is predominately a sequential process. It cannot be easily rushed. Initial production programs should recognize the sequential character of learning and exercise opportunities to enhance learning via a properly phased production schedule. Timing for program phasing should reflect the time required to implement a complex, multi-billion dollar synfuel project. A practical timeframe for design, construction, and initial operation indicates a minimum time of about five to seven years between phases. Time required for procedural approvals in today's regulatory environment could add several years to this minimum spacing. Even with expedited projects, operating experience to support assessments for a second phase are not likely to be available until fully five to seven years after initial production projects get started. Production schedules that do not respond to these technological realities may repeat mistakes and forego important opportunities for learning.

Diversity Fosters Learning

Both selection learning and technology improvement learning are fostered by some diversity for the technologies and coal resource bases employed as part of an initial production effort. Learning benefits accrue primarily when practical experience is applied to improve technologies and decisions for future deployment efforts.

Consequently, experience with a relevant range of technologies and coals is important for potential surge deployment assessments that must determine both *which* technologies to deploy and *how* to build improved versions. Initial production efforts should recognize the value of a diverse experience base that can improve selection information, enhance the general applicability of initial production experience to future deployment efforts, and take advantage of the significant potential for cross-technology learning.

Information Is a Key to Learning

Information collection, sharing, and transfer are important to the realization of potential benefits from both selection learning and technology improvement learning. Special attention should be given to documenting the information outputs of initial production efforts. Although common commercial practice does not naturally foster such information development and exchange, important information outputs may be available in forms that do not transgress on important proprietary positions or commercial incentives. Program design and implementation should assess, structure, and foster the generation, documentation, and transfer of information relevant to improving future decisions.

Smaller Projects and Programs May Foster Learning

In general, more experience should yield more learning. However, the similarity and limited number of existing commercially-proven technologies for coal-based synfuels make it likely that large production levels would be achieved by building several plants using essentially the same technology. The learning value of multiple, simultaneous projects using the same basic technology is limited.

A smaller-volume initial production phase comprised of some (but not necessarily all) relatively small commercial projects may well foster learning more than a larger volume effort comprised of full-size facilities. More smaller projects provide more comparisons for selection learning. Smaller projects in a smaller-volume first phase may procede more expeditiously, providing earlier initial practical experience and reducing the hiatus period before a major increment of learning becomes available. Relatively small commercial plants provide the same technology improvement information as larger plants. Moreover, they can have significant advantages if technology innovation is to be encouraged by using riskier, less proven technologies, or if experimentation with operating conditions to identify technology improvements is to be encouraged with initial commercial projects.

The Importance of Defining Priorities Among Objectives

Learning is but one of several concurrent benefits and purposes of initial synfuels production efforts. Although many approaches for enhancing learning are possible, other program objectives must be considered in developing a cost-effective program. Some program design and implementation prescriptions for enhancing learning may also serve other objectives, but conflicts and tradeoffs may arise. This chapter has noted stylized issues for an emphasis on learning and suggested directions that should be taken for further, more detailed analysis of the development of cost-effective program design structures and provisions.

Ultimately, preferences and appropriate choices will depend on the priorities established among various purposes. Preferences among objectives have important implications for program design and implementation and should be explicitly and carefully considered by responsible decisionmakers.

NOTES

1. This frequently occurs with new technologies. Cost estimates show a clear tendency to increase as a project comes closer to reality and full complexities are realized. The effect is qualitatively and empirically reviewed in Merrow, Chapel, and Worthing (1979) and Merrow, Phillips, and Myers (1981).
2. An example of such a generic technology might be Lurgi gasification followed by some process to convert synthesis gas to another product. For

PROGRAM DESIGN ISSUES FOR ENHANCING LEARNING   273

example, given engineering similarities, we might expect Lurgi gasification followed by a *methanol* production step to have costs quite similar to Lurgi gasification followed by a *methane* production step. Cost distributions in such cases could be so similar as to preclude ready differentiation. Cost in this context should reflect the varying end-use costs or form value for different synthetic fuel products.

3. Given a propensity to adopt proven technology for capital-intensive projects, the technological bias could never be reversed or even discovered. For example, some writers have suggested that U.S. commercial development of nuclear reactors foreclosed, without adequate basis, the development of alternative nuclear reactor designs. See, for example, Bupp and Derian (1978).

4. The study was the Rand Corporation's Pioneer Plants Study. See Merrow, Phillips, and Myers (1981).

5. For a good overview article summarizing the learning curve concept along with some empirical discussion, see Bodde (1976) or Hirschman (1964). For discussions with a more empirical emphasis, see Mooz (1978, 1979), Ostwald and Reisdorf (1979), Joskow and Rozanski (1979). A good review of learning effects for capital intensive oil shale technologies may be found in Merrow (1978). Some of Merrow's ongoing empirical work is directly relevant to rates of learning with innovative chemical process technologies similar to technology for synthetic fuels from coal.

6. The descriptions in this section are drawn from the general experience of the author as well as from interviews with several experienced personnel of the architect/engineering industry. General discussions with personnel at three different architect/engineering firms were held as follows: Badger America, Inc., William Henze, Vice President for Synthetic Fuels, on February 14, 1980, and April 12, 1980; Fluor Engineers and Constructors, Dana Lee, Manager, Government Relations, and others, May 1, 1980; Stearns-Rogers and Company, Edward E. Ives, Manager for Synthetic Fuels on May 4, 1980. Background written materials describing learning phenomena include Merrow (1978), Enos (1962), Bodde (1976), and Hirschman (1964).

7. The discussion of project phases for description of learning reflects information and descriptions that are commonly applied. However, the descriptions and terminology applied here draw heavily from the descriptions in Merrow, Chapel, and Worthing (1979: 55-61).

8. These observations on the contribution of design experience alone to learning are drawn largely from discussions with AE&C personnel, especially those at Badger America, Inc., in Cambridge, Massachusetts. The general notion that operating experience is critical to learning for innovative process plants is widely accepted in the chemical engineering profession. A major empirical study for the Department of Energy shows that performance of innovative process plants is related to the number of new or undemon-

strated process steps and the extent to which a process has been characterized by its heat and mass balance equations. Actual operating experience modifies both of these factors in directions that reduce costs. See Merrow, Phillips, and Myers (1981).

9. Discussions with AE&C firms confirmed that many process engineering firms learn a great deal from start-up experience and place emphasis on having design personnel at the start-up site to learn directly from the start-up information.
10. The propensity for learning during start-up operations or process upsets may be associated with the combined presence of design, scientific, and operating personnel. The process is described in Hirschman (1964).
11. See Mooz (1978, 1979) for background on this point as applied to light water nuclear reactor technology.
12. Construction on pioneer plants is often deferred until a definitive engineering design is completed. See Merrow, Chapel, and Worthing (1979: 59).
13. Attempts to short-circuit design and engineering phases for a pioneering facility may cause significant problems in the construction phase. (personal communication between author and E.W. Merrow, June 1980). The effect is also noted in Merrow, Phillips, and Myers (1981) and in Brusher (1979). Brusher describes the importance of clear and complete project definition for efficient planning and smooth project execution.
14. After some point in the design and construction of a project, the incorporation of new design improvements based on experience from other facilities becomes very difficult and costly to implement. Important elements of plant design may be frozen at some point, even though improvements could have been readily implemented earlier in the progress of a project. The point at which a design becomes substantially frozen depends on the particular project and the costs of rescheduling or redesigning relative to the cost savings available from incorporating the improvement.
15. However, expedited plants still may be desirable, since they will reduce the hiatus period during which any major surge deployment effort must be based on very limited practical experience.
16. Reported by David Goerz of Bechtel National, Inc. during informal telephone conversation with author on April 25, 1980.
17. This observation was made by personnel from Badger America, Inc., during information interviews with the author in Cambridge, Massachusetts, on April 10, 1980.
18. This common industrial practice was described by all the AE&C firms interviewed.

9 BUILDING DEPLOYMENT CAPABILITY
Strategic Issues and Analytical Approaches

Many constraints could combine to limit the number of synfuel plants that can be simultaneously deployed in a surge deployment effort. Initial production experience can build industry infrastructure, relax some of these constraints, and expand future deployment capability. Such enhanced deployment capabilities are an important source of strategic insurance benefits from initial synfuel projects. This chapter reviews major program design and implementation issues for enhancing infrastructure development.

As with learning benefits, approaches for building deployment capability are identified by characterizing the linkages between practical production experience and the development of an industry infrastructure. However, each of the enormous number of inputs required to build a synfuels plant raises many issues for the process by which deployment capability increases. Constraints are like the layers of an onion. When constraints for one input are relaxed, another constraint becomes limiting at a slightly higher level of deployment. We will not attempt to characterize and assess the numerous complex issues. Rather, this chapter focuses on illustrating a basic analytical approach for identifying strategic deployability issues and uses the characterization of linkages to suggest possible ways to do better. Prudent, cost-effective program design will require further assessment of many ideas and issues suggested by this review.

Deployment constraints and infrastructure issues are critical to the design of cost-effective initial production programs. It is important to recall that many of the phenomena underlying the potential benefits of enhanced deployability are also related to the potential for added costs for initial production efforts due to factor cost inflation. In a sense, deployability benefits and factor cost inflation are two sides of the same coin. The discussions in Parts I and II of this study have demonstrated the size and relative importance of both the benefits and costs associated with various infrastructure effects.

Unfortunately, as described in Chapter 3, quantification of infrastructure effects and factor cost inflation is a complex and speculative task. The likelihood of deployment constraints at various levels depends on developments outside the synthetic fuels industry. The level of factor cost inflation is also uncertain because it depends on how suppliers of the material, labor, and skilled service inputs to the synthetic fuel deployment process respond to substantial increases in both demand for their goods and services (and the resulting price levels). Despite these quantitative difficulties, provisions for improving the cost-effectiveness of initial production efforts can be identified by attention to the qualitative linkages that help characterize both existing deployment constraints and key opportunities for building infrastructure to expand future deployment capabilities. Cost-effective program design provisions could emphasize either building strategic infrastructure for the future or reducing unnecessary added costs from factor cost inflation during initial production efforts.

Recall that the sources of enhanced deployability addressed in this study stem from relaxing constraints on inputs that limit the number of simultaneous projects possible in a surge deployment effort. The benefits from more and sooner synfuel production could also be obtained by relaxing the procedural constraints that prolong the time required for project completion rather than by increasing the number of simultaneous projects that are possible. Efforts to reduce permitting and approval processes or to telescope planning and scheduling lead-times could reduce project completion times and advance the time at which resources from one project are free to be applied to another project. Relaxing procedural constraints via efforts such as site banking or expedited permitting could be an important avenue toward the benefits of more aggressive synfuels growth potential.[1] However, procedural constraints and the means of relaxing them have been explicitly excluded from this study. Broader assessments should compare provisions for relaxing procedural constraints with other means of enhancing the rate and extent of potential industry expansion.

This chapter reviews issues for enhancing infrastructure development to increase the number of simultaneous projects possible. Inputs to the synfuels deployment process are grouped into two major categories. "Service inputs" include all the "soft" or non-hardware items required to undertake a synfuels project. Key service inputs include architect/engineer design services, construction management services, and even the managerial and entrepreneurial capabilities of the project sponsors who own and operate a plant. Construction labor inputs, especially specialized construction skills, are also important service inputs. "Hardware inputs" include the tangible equipment, components, and materials required for the construction of a synthetic fuels facility. Examples of potentially constraining hardware items include heat exchangers, specialized valves, and compressors.

Separate sections of this chapter address each of these major categories of inputs. First, however, an initial section outlines an approach for identifying priority constraints and assessing the linkages between initial production experience and the relaxation of key constraints.

A STRATEGIC APPROACH FOR ASSESSING DEPLOYABILITY ISSUES

Each category of inputs to the synthetic fuel deployment process will include items likely to become constraints at some deployment level, but not all constraints are equally important. We should recognize that enhanced deployment capability is a strategic insurance capability that becomes important primarily in the event of an all-out or emergency surge deployment effort.

This "strategic capability as insurance" view of potential deployment constraints allows a step-wise analytical approach to focus attention on a limited number of potential constraints with particular strategic importance. The first step of this approach involves identifying the constraints likely to limit various levels of deployment. The second step describes the process by which key constraints may be relaxed. After these two steps, it should be possible to separate potentially important constraints into two distinct categories. "Long lead-time" constraints are those that require significant time to relax. Such constraints can be relaxed only via strategic investments *before* a surge deployment effort is initiated. "Short lead-time" constraints are those that can be effectively relaxed, albeit perhaps with major efforts, *after* a surge deployment effort has been initiated.

For example, the availability of qualified and experienced synthetic fuel project management teams may be a long lead-time constraint, because the experience required to become fully capable can be acquired only over the course of a seven-year synthetic fuel project. In contrast, the availability of qualified pipefitters is more of a short lead-time constraint, because additional craftsmen can be trained on the job over the course of several months.

Strategic insurance investments are made in order to prepare for the possibility of an unfortunate event. If events do not occur, the insurance investments yield little return (unless they affect the probability for events). Where possible, investments in infrastructure development should be deferred until after an event has made major capacity expansion clearly warranted. In general, investments to relax short lead-time deployment constraints can and should be deferred until a surge deployment effort has begun. Accordingly, strategic insurance investments in enhanced synthetic fuel deployment capability should stress long lead-time constraints. Moreover, particular emphasis should be placed on the subset of long lead-time constraints that can be relaxed as a result of the economic activity and experience provided by initial production projects. Existing long lead-time constraints can limit the practicable size of an initial production effort.

Separating Long from Short Lead-Time Concerns

The first major analysis step in a strategic assessment of deployability issues characterizes inputs as long lead-time concerns or short lead-time concerns. The analysis should examine both the process for relaxing constraints and the timing required for that process. Some potential constraints can be effectively relaxed during implementation of the initial production program or surge deployment effort; others cannot. For example, if trained pipewelders were likely to become a problem during construction of a synfuel plant, project management could institute training programs to equip available general labor with the requisite skills. Such efforts to mitigate or avoid constraints may require special programs, such as training agreements with craft unions. However, the time or other resources required to train personnel should not present a serious limitation.

Once a particular input or issue has been properly identified as a short lead-time concern, extensive analysis and investment to relax such constraints may not be necessary or worthwhile until all-out surge deployment effort is underway. Further assessment and characterization of measures to relax short lead-time constraints are not high priority concerns for program design emphasizing strategic investments in future surge deployment capabilities. In contrast, constraints for long lead-time inputs merit additional attention and analysis.

Understanding How Production Experience Relaxes Constraints

Identifying long lead-time constraints is only an initial step. Important long lead-time deployability constraints should be qualitatively and quantitatively characterized. Moreover, the process by which an initial production experience may or may not relax long lead-time constraints should be considered.

Basically, initial synfuels production projects generate activity that exercises the engineering/industrial system for synthetic fuels deployment. The practical experience builds infrastructure that may increase capacity or improve efficiency for key inputs required to deploy synfuels plants. The activity and experience provided via an initial production program may or may not relax a particular long lead-time constraint.

Suppose that a surge deployment effort would strain production capacity (both domestically and worldwide) for specialized compressors. An initial synthetic fuels production program, depending on its scale and duration, may not necessarily lead to expansion in the production capacity for that long lead-time input. Although the initial production efforts would generate additional demand for the compressors, this additional demand might simply use up existing excess production capacity or cause price increases rather than induce suppliers to expand production capacity. Historically, equipment manufacturers supplying the energy industry build new production capacity primarily in response to assured, continuing demand (Bechtel 1979: 4-25). Therefore, the scale and duration of an initial production program would have a direct effect on whether or not equipment suppliers expand production capability. Such expansions in production capacity for selected equipment items could be a key source of enhanced deployment capabilities. One should ask, however, whether such infrastructure development could be obtained more cost-effectively by initial production projects (which generate the demand which engenders capacity expansion) or by investments more directly targeted toward building surge production capacity for key, potentially constraining, long lead-time equipment items.

Capabilities for other potentially constraining, long lead-time inputs, however, could be substantially improved via experience from initial synfuel projects. For example, suppose that deployment efforts were limited by the availability of design, construction, and project management teams experienced with implementing a multi-billion dollar synfuel project. The experience that increases the number of organizations and personnel capable of undertaking a synthetic fuels project can be acquired only via practical participation with project implementation.

Program design and implementation provisions oriented toward increasing infrastructure development should be based on assessments similar to the examples above. The approach reflects a logical progression that assesses which inputs may become binding, which inputs are truly long lead-time concerns, and which constraining, long lead-time concerns can be practicably relaxed via the activity and experience generated by initial synthetic fuel projects.

The application of such systematic analysis to the many inputs required for the synfuels deployment process is beyond the scope of this study. However, the general analytical approach can be useful for illuminating some key program design issues associated with each of the major categories of inputs—services and hardware goods.

The following sections consider each of these two major categories with emphasis on applying the step-wise strategic analysis to a few key issues. This characterization suggests program design provisions that may enhance future deployment capability or reduce factor cost inflation for initial production efforts. Later, more detailed studies will be needed to assess whether particular provisions are practicable or cost-effective.

SERVICE INPUTS

About 40 percent of the construction costs for a synthetic fuel plant are intangible services rather than materials and equipment (Bechtel 1979: 2-5). Important categories of such service inputs include technical services for project definition and plant design, management services, supervisory services, skilled labor, and general labor. The firm that owns and operates a synthetic fuel project also provides management, technical, and financial services to the project. Assessments of synthetic fuel deployment requirements often note potential constraints arising for two different service inputs: architect, engineering, and construction management services (AE&C) and specialized construction labor skills.

These two different service inputs are useful for illustrating application of the general analytical approach for strategic investments in enhanced deployment capability. Potential constraints for architect, engineering, and construction management receive substantial attention, because development of AE&C capabilities is clearly a long lead-time issue. Review of constraints arising for construction labor services is much briefer, since those inputs are primarily short lead-time concerns.

AE&C Services

An important service input constraint on synthetic fuel deployment capability arises from potential limits on the number of architect, engineering, and construction (AE&C) firms that are well-qualified to undertake the design or to manage the construction of a complex, multi-billion dollar synfuel project. Assessments of the magnitude of potential AE&C constraints vary. Edward Merrow of the Rand Corporation has suggested that only five to ten AE&C firms are presently capable of undertaking a coal-based synthetic fuels project

(Merrow and Worthing 1979). Of this limited number of firms, only a few could undertake more than one project at the same time. Although some standardization of design or subcontracting techniques could relax any AE&C constraint, such techniques have limits and certainly would introduce added costs.[2] Moreover, project management, construction management, and procurement services are important AE&C functions that consume limited resources, even if designs are standardized.[3]

Other assessments of AE&C constraints emphasize limits on trained talent from some engineering disciplines rather than limits on the number of synfuel-capable firms. Cameron Engineers (1979) observes that only twenty-one AE&C firms have total annual business volumes greater than $400 million annually, an amount substantially less than the cost of a single, commercial-scale synthetic fuels facility. Although the Cameron Engineers report goes on to note that joint ventures among AE&C firms are possible, it also cites the "critical" demand for "managers, engineers, draftsmen, and other professional people to plan and design the new facilities." Two recent studies of supply issues associated with synthetic fuels deployment in the 1980s specifically identify the supply of chemical engineers as a key, potentially constrained input (Bechtel 1979: 1-20; Dineen, Merrow, and Mooz 1981).

A full explication of the nature and extent of potential deployment constraints arising from the availability of capable AE&C services requires detailed scenario-specific assessments. Constraints identified and quantitatively estimated under one scenario might be altered if the scenario changed or if one of various techniques for relaxing potential constraints were employed. For example, AE&C constraints may depend on assumptions about the level of new process engineering and construction in the chemical industry. Changes in the synthetic fuel technology mix could change the nature of constraints.[4] Nevertheless, qualitative review of potential constraints associated with AE&C services suggests program design issues and illustrates analytical approaches for improving the cost-effectiveness of initial synthetic fuel production efforts.

Is the AE&C Constraint a Long-Lead-Time Concern? Although it is likely that, at some level of deployment, limitations for AE&C services could become a binding constraint, further description of the exact nature of the constraint is needed to determine whether it is a long lead-time constraint or not. In particular, some assessments suggest that the primary source of the AE&C constraint is experienced managerial teams and established organizational systems for efficiently undertaking the design and construction of a complex, multi-billion dollar synfuels project.[5]

If organizational capabilities and management teams experienced with very large "mega-projects" are the key to the AE&C constraint, the next analytical task step involves characterizing the linkages by which initial synthetic fuel production experience can increase future AE&C capacity. If the number of AE&C firms with mega-project capabilities is not increased following an initial production effort, then the initial production experience does little to relax any AE&C constraint. If, however, the number or capacity of synfuels-capable AE&C firms increases via initial synthetic fuels production experience, the nation's posture for responding to surge deployment events may be substantially improved.

How Is AE&C Capability Increased? The particular description or model for increasing AE&C capacity has implications for both cost-effective program structure and the magnitude of expected deployability benefits.

Three alternative models can illustrate possible mechanisms whereby initial production experience increases the number and capacity of AE&C firms that are well qualified to take on a full scale synfuels project. In the first model, a firm that is already synthetics-capable could, via expanding experience, increase its ability to efficiently take on multiple projects at the same time. According to a second model, experienced personnel from a synfuel-capable firm could move to an underqualified AE&C firm (or establish a new firm), thus substantially increasing that firm's capabilities for successfully implementing a major synthetic fuels project. Under a third model, a presently marginally-qualified AE&C firm, by virtue of the experience of performing an initial project, could become fully established as a "synthetics-capable" firm. There are many possible variants and combinations of these stylized models for the process by which AE&C capability grows.

The strength of the linkage between initial production experience and expanding AE&C capability varies with the applicable model or description. For example, under the first model, AE&C capacity is unlikely to increase in direct proportion to the number of initial production projects undertaken. Experience with one project may not necessarily double organizational capabilities for subsequent projects. Under the second model, providing experience to management teams may not necessarily lead to increased AE&C capacity, because personnel may not readily transfer to other firms, because older project managers may retire, or even because experienced personnel may not be fully effective in firms without sufficiently well-established organizational systems for mega-project development, design, and management. Under the third model, experience with one synfuel project may well make a firm capable of handling an individual project in subsequent deployment efforts, but may not necessarily make such a newly synfuels-capable firm competent to undertake efficiently the multiple projects that could be handled by other synfuel-capable firms with a broader experience base.

At this point, we do not know which model (or combinations and variants) most aptly describes the process by which an initial synfuels production experience expands AE&C-related deployment capabilities.[6] However, improved descriptions of AE&C constraints can suggest and clarify issues for cost-effective program design. For example, if the predominant importance of the third model for increasing AE&C capabilities is stressed, then program design might emphasize the involvement of AE&C firms that do not yet have the demonstrated experience at the mega-project scale necessary to make them clearly synthetics-capable. Initial production experience, perhaps with smaller than average projects, could train those firms so that they would become capable of efficiently undertaking projects in future surge deployment efforts. Such a program objective could be accomplished, for example, by having an initial production program large enough to necessitate the lead involvement, for at least a few projects, of some presently marginally-qualified AE&C firms. Alternatively, program implementation for a subset of projects could prefer or require the prime involvement of AE&C firms not quite synthetics-capable as a condition of the proposal or contract that provided program financial incentives.

Evaluating Cost-Effectiveness for a Particular Provision. Such program design provisions could enhance the development of the engineering and project management infrastructure to support future aggressive expansion. However, cost-effective program design requires the additional step of comparing the incremental costs of any program design provisions to the incremental benefits. For example, lead involvement of an AE&C firm not fully experienced with multibillion dollar projects might increase the expected cost of a synfuels project by $5 per barrel. The incremental cost of the program design provision resulting in such a $5 per barrel increase for that one project in order to "train" an additional AE&C firm would have a net present value of $0.8 billion (1980$) for a full size project.[7] (Incremental costs would be less for a smaller project, scaling in direct proportion to production volume.)

This certain incremental cost should be compared to the *expected* incremental benefits that stem from being able to build an additional synthetic fuel plant in a surge deployment effort. Expected incremental benefits may be estimated by multiplying the value of an additional plant by the probability of the occurrence of a surge deployment event.

The net comparison of the costs and benefits for a particular program design provision will depend on the probabilities for the surge deployment events. The greater the incremental insurance payoffs as compared to the incremental premium, the lower the probability of surge deployment events for which a particular program design provision remains cost-effective. Thus, the value of special program design provisions to enhance infrastructure development for future deployment capabilities depends on individual expectations for oil prices, synfuel costs, and the probabilities for surge deployment events.

Assessing Other Issues for Relaxing AE&C Constraints. Constraints for AE&C capabilities are likely to be an important concern for both initial production programs and future surge deployment efforts. The notion of using initial production efforts to train marginally qualified AE&C firms is but one of many possible approaches for relaxing potential AE&C constraints. Descriptions of the AE&C constraint suggest other issues that merit further assessment, including:

- The costs and merits of special training programs for chemical engineers needed for synfuel project design;
- Prospects for project selection that reduce requirements for AE&C services by favoring technologies requiring less design effort;
- The advantages and disadvantages of standardized designs to stretch limited AE&C capabilities;

- The development, demonstration, and evaluation of other measures for leveraging limited AE&C capabilities, such as division of labor between different firms for various parts of a project such as design/engineering and construction management; and
- The possible role for owner/operator firms that have substantial in-house design and engineering capabilities.
- The potential atrophy of people-based capabilities over time.

The list of issues related to potential AE&C constraints could go on. Improved understanding—based on careful descriptions of how AE&C capability may limit deployment capability and how any constraint may be relaxed—is necessary to assess the importance of various issues and the merits of particular program design provisions. Ultimately, appropriate choices concerning particular provisions will depend on individual judgments.

The systematic analytical approach illustrated for AE&C constraints can be applied to other issues and constraints on synfuels deployability. Identification and assessment of cost-effective program design provisions has three basic elements: (1) describing which constraints are likely to be severe limitations and the process by which such constraints may be relaxed via the experience and activity generated by initial synthetic fuel production projects; (2) identifying program design provisions, based on these descriptions, which can enhance infrastructure development for future deployment or avoid factor cost inflation for the initial production effort itself; and (3) comparing any certain incremental costs of such program design provisions with expected (probability-weighted) incremental benefits.

Application of this general approach will not provide easy or obvious answers. The issues are complex. However, systematic thinking that recognizes the dynamics and key role for strategic infrastructure development should help identify cost-effective approaches for doing better with respect to AE&C constraints or other long lead-time concerns for synfuels deployment capability.

Construction Labor

Studies of synthetic fuel deployment often cite potential constraints for the skilled construction labor required to build many synthetic fuel plants simultaneously. Even the availability of general unskilled labor can be problematic for projects located in sparsely populated regions. More importantly, regional or nationwide shortages could arise during a major synfuels effort for some specialized construction trades (such as pipefitters, boilermakers, welders, and electricians).[8] Limited availability of qualified high-pressure welders, for example, could cause labor scheduling difficulties that would effectively increase the costs of all plants under construction and limit the maximum size of deployment efforts.

The key strategic issue for this potential deployment constraint concerns whether skilled labor inputs are a long lead-time or short lead-time constraint. Clearly, training programs to increase the supply of qualified personnel could relax any potential deployment constraints. However, the time required to train additional craftsmen for key construction services is a key issue.

If training can be accomplished in a short period of time relative to the construction period for a synthetic fuel plant, then labor constraints are short lead-time concerns. Strategic attention and investments for capacity expansion for key labor skills in advance of a surge deployment effort may not be necessary or appropriate. However, if development of labor skills requires several years of experience and training, then a strategically designed initial synfuels production effort might include provisions explicitly oriented toward increasing the supply of experienced persons who are well qualified with those particular specialized labor skills which are likely to limit a surge deployment effort.

The time required to expand the availability of skilled construction workers is affected by the policies of the various construction trade unions. Many crafts unions have extensive training and apprenticeship requirements. Union policies and practices can have substantial influence on both the effective labor supply and the rate at which it can be increased. Typically, many crafts unions are hesitant to expand apprenticeship programs and union membership in response to transient peaks in demand.[9] Consequently, although there may be, in theory, adequate opportunity to train skilled personnel to support major deployment efforts, union practices could operate to inhibit the expeditious training of required skilled craftsmen. Further analysis is required to assess whether such practices would occur and act to constrain either initial production projects or possible future surge deployment efforts.

Overall, appropriate planning efforts should enable the supply of skilled labor to be properly considered a short lead-time concern. Advance strategic investments are not needed, since required labor supply can be developed after a surge deployment effort has been initiated. The National Constructors Association notes that basic training for many construction labor skills, such as welding, requires only several months; further training can take place on the job.[10] Skilled laborers for the Sasol-II project in South Africa were trained as part of well-planned on-site training programs.[11] Bechtel National, a major AE&C firm, reports that "a person with limited welding experience can be upgraded to nuclear quality in six to twelve weeks" (Bechtel 1979: 4-23). Clearly, while such training programs require effort and attention (and some increased construction costs), construction labor is not generally not a long lead-time concern.

However, if some skilled labor requirements were determined to be long lead-time inputs, other steps in a strategic analytical approach should be applied. For instance, by what process does the activity during an initial production program relax any constraint? What practical program design provisions are suggested by a description of the linkages? Should synthetic fuel incentive contracts include a provision for specialized labor training programs as part of project implementation? As usual, cost-effective program design requires comparing any additional costs of particular program design provisions with the expected increase in benefits.

HARDWARE INPUTS: EQUIPMENT, MATERIAL, AND COMPONENTS

About half of the construction cost of a synthetic fuel facility represents tangible goods such as material, equipment, and components (Bechtel 1979: 2-15). Some of these goods, such as cement, are best produced near the plant site, but the majority are procured on national or international markets and transported to the plant site.[12]

The assessment of potential shortages and constraints for equipment and materials is problematic because a shortage depends on the situation for both demand and supply. Material requirements for a given synthetic fuels deployment scenario can be reasonably well-specified. However, economy-wide demand levels depend on industrial needs and economic activities from sectors of the economy other than synfuels.

Even if total demand is well-specified, total supplies depend on production capacity and supply availability from both domestic and international suppliers. Such assessments of the supply side for hardware constraints require sophisticated, often subjective, judgments of both existing spare production capacity and the likely response of producers to higher levels of demand (as well as prices and profits). For example, as demand conditions warrant, suppliers can expand production either by building additional plant capacity or by changing to multiple shift operations in existing facilities.[13]

Defining Potentially Constrained Items

Despite these assessment difficulties, studies of synthetic fuel deployment commonly cite a number of material, equipment, and component items as potential constraints. For example, one assessment of deployment requirements by the Flour Corporation (1979) reported that a deployment effort of 1.5 million barrels per day (including 500,000 barrels per day of shale oil) would require more than 25 percent of the 1978 U.S. manufacturing capacity for compressors, pumps, fired heaters and boilers, and valves. However, the Flour study also observed that equipment vendors generally thought that, given adequate lead-time, additional production capacity could be brought on line to supply materials requirements with "very few problems" (Flour 1979: 38). The Bechtel study (Bechtel 1979) examined potential impediments to a deployment of one million barrels per day of coal liquids production capacity in the 1980s; it addressed material requirements and supplier responses in some detail. Table 9-1 displays the major materials and equipment requirements that were identified as likely constraints.

Assessing Strategic Investments To Relax Against Hardware Constraints

Table 9-1 shows that a major synfuels deployment could require substantial shares of production capacity for several categories of materials and equipment. Limited availability of essential hardware items could complicate the construction process causing delays and cost increases. The maximum size of a surge deployment effort could be limited by practical logistical concerns.

Table 9-1. Materials and Equipment Items Potentially Constraining a Major Synfuels Deployment Effort (for deployment scenario of 1 MMBD by 1990, 3 MMBD by 2000).[a]

Category	Peak Year Annual Requirements as Percentage of U.S. Production Capacity
Draglines	88%
Heat exchangers	74%
Centrifugal compressors[b]	18%
Pressure vessels[c]	12%
Valves[d]	8%

[a]Table adapted from Bechtel National, Inc., 1979, "Production of Synthetic Liquids from Coal: 1980-2000," Table 1-9, p. 1-27. Two other potentially critical items noted by Bechtel were pumps and drivers and chromium. Both of these items had peak year annual requirements less than 5 percent of U.S. production capacity, 4 percent and 3 percent respectively.

[b]Less than 10,000 horsepower.

[c]Both greater than 4-inch wall and 1.5- to 4-inch wall.

[d]Alloy and stainless steel.

However, it is not at all clear that potential constraints for material and equipment are long lead-time concerns meriting strategic attention before a surge deployment effort is initiated. Assessments of the merits of strategic investments in relaxing constraints for any particular component require component-specific characterization of the time required to build infrastructure and expand production capacity.

For example, if high performance valves are a potential constraint, industry-specific analysis should assess the time required to expand valve production capacity. Such analysis should include the complete system for production of a potentially constrained item. If valve manufacture requires high quality forgings as a raw material, then the time required to expand supplies of such high quality forgings should also be reflected in the analysis. A hardware item becomes a long lead-time concern primarily if the time required to expand supplies exceeds the time at which the component is needed in the construction schedule for a synfuels plant. Of course, such construction schedules can be adjusted (presumably at some cost) to accommodate delays in availability of key components.

For those items identified as long lead-time concerns, further analysis should examine how initial synfuel production activities may or may not relax a given constraint. Component suppliers (like trade unions) may not necessarily expand production capacity in response to temporary surges in demand. Consequently, a program for steadily increasing production levels may lead to more permanent infrastructure development for key inputs (such as building new capacity rather than only moving temporarily to multiple shift operations) than a production program concentrated into only one or two phases. A phased initial production effort with surges of demand may not be as effective at building continuing infrastructure as a program committing reliably to steady industry growth. If hardware constraints are an important, long lead-time concern, the advantages of a phased program for learning may trade off with the infrastructure development objective.

Characterization of the process for expanding supplies of key long lead-time equipment items may suggest program design provisions that could enhance infrastructure development. For example, incentive contracts for initial projects could encourage use of new production capacity (e.g., valve factories) for long lead-time items. Alternatively, it could be more cost-effective to build infrastructure by direct investment in excess or surge capacity for that item rather than by inducing capacity expansion via the added demand of initial synfuels projects. Such an approach recognizes the strategic character of investments in infrastructure for enhanced deployability and has been applied by the military for selected needs. Rather than attempting to relax any "compressor constraint" suggested by Table 9-1 via a large number of initial projects, it could be more cost-effective to establish a "strategic compressor reserve." Of course, such advance strategic investments would not be warranted unless availability of compressors was clearly a long lead-time constraint. Again, cost-effective program design requires that the incremental costs of such efforts be compared to expected incremental benefits.

Most Hardware Constraints Are Short Lead-Time Issues

Although detailed, component-specific implementation of the analytical approach outlined above has not been attempted, a general review of several studies of deployment constraints indicates that most (if not all) potential material and equipment requirements are likely to be short lead-time concerns (Bechtel 1979; Flour 1979; Synfuels Interagency Task Force 1975). Given adequate time, credible committments, and market incentives, producers of material and components can expand supplies to meet the requirements of aggressive synfuels capacity expansion. The Bechtel study concludes with the statement that "availability of material and equipment is not expected to present a serious obstacle" to deployment of one million barrels per day of coal synthetics by 1990 (Bechtel 1979: 1–23). The Bechtel conclusion was supported by observations: that only a small share of total U.S. manufacturing capacity is required to produce many components for synfuels plants; that excess capacity exists for the manufacture of many components; that manufacturers are able to rapidly expand production in response to developing demand; and that worldwide procurement of component requirements is possible (Bechtel 1979). Most hardware inputs are short lead-time or tactical concerns that do not require strategic insurance investments in initial synfuels production experience in order to make more plants possible in a surge deployment effort.[14]

DOING BETTER FOR INFRASTRUCTURE DEVELOPMENT

Issues for "doing better" for infrastructure development are numerous and complex. Definitive assessment of cost-effective approaches require further analysis. The general review and strategic approach illustrated in this chapter, however, suggest several key concepts and issues.

Long Lead-Time Issues Provide Focus for Program Design

"Infrastructure development" is a summary term that refers to expanding supplies and capabilities for the numerous inputs required to build synfuels plants; general experience with synfuels should build infrastructure. However, given a strategic insurance view of the purpose of infrastructure development, emphasis on long lead-time constraints usefully limits the number of inputs and issues that must be addressed. A priority task for cost-effective program design is the proper characterization of various constraints as either long or short lead-time concerns. Strategic investments for future deployment capability should be focused on long lead-time constraints. Moreover, good characterization of existing long lead-time constraints can help define the appropriate size for an initial production effort. Initial production programs that seriously strain the existing capacity for long lead-time inputs are likely to encounter substantial factor cost inflation.

AE&C Capabilities Are a Key Long Lead-Time Input

A basic review of the major classes of inputs required for synfuel deployment suggests that constraints for experienced, synfuels-capable architect, engineering, and construction (AE&C) services are a key long lead-time concern. Many other inputs, such as specialized construction labor or equipment items, are important short lead-time concerns that can be handled as a tactical matter as part of an initial production program or after future surge deployment efforts have begun.[15]

If AE&C capabilities are a key long lead-time constraint, the existing capabilities of the AE&C industry may provide strong guidance for the practicable number of an initial synfuels projects. Future deployment capability may be enhanced by program design and implementation provisions that seek to expand the number and capacity of AE&C firms that are capable of efficiently undertaking a synfuels project. Improved understanding of potential AE&C constraints and the process for relaxing the constraints should help suggest cost-effective program design provisions.

Payoffs from Infrastructure Development Depend on Expectations for Our Energy Future

Efforts to build infrastructure for future deployment capability are strategic insurance investments providing benefits in special scenarios. The merits of increased program size or special provisions designed to accent infrastructure development depend on the likelihood of events that could make it valuable to deploy synfuels aggressively at maximum practicable rates. Thus, the cost-effectiveness of special program emphasis or provisions to enhance infrastructure development depend on individual views of our uncertain energy future.

Issues for "Doing Better"

Unfortunately, there are few clear-cut answers for the "How to do better?" question concerning to deployability benefits. However, it is clear that building strategic deployment capability involves more than simply increasing the size of an initial synfuels production effort. Clarifying program purposes and asking questions about the relationship between how we proceed with initial synfuels production efforts and how our national posture is thus improved are important steps toward doing better.

NOTES

1. Relaxation of procedural constraints was a major focus of the Energy Mobilization Board proposal in 1979. See White House (1979).
2. Several studies cite service constraints in these areas. Bechtel National, Inc. (1979) cites potential engineering manpower constraints as well as specialized construction labor constraints. Cameron Engineers (1979) notes potential constraints on the number of design and construction firms capable of handling a synfuels project. The Cameron Engineers study also notes potential construction labor constraints in some localities. Several Rand Corporation studies (Merrow and Worthing 1979; Dinneen, Merrow and Mooz 1981) note potential constraints arising from capacity in the architect, engineering, and construction (AE&C) industry as well as possible shortages in some key construction labor skills or equipment items. This author's interviews with personnel in the AE&C industry have indicated that another possible constraint on synthetic fuels deployability is the number of well-qualified owning/operating firms. There are a limited number of firms with the organizational wherewithal to successfully undertake a single project representing a multi-billion dollar activity and exposure. Many oil companies as well as some chemicals and minerals companies are familiar with projects at the investment scale of synthetic fuels. In order to relax this potential constraint, an initial production program

might attempt to expand the base of entrepreneur/owner firms capable of attempting future synthetic fuels projects. For example, firms with strong in-house engineering or construction management capabilities might be preferentially involved in order to leverage the supply of AE&C services. Also, some persons might be interested in using an initial production program to enhance the competitive structure of an evolving synthetic fuels industry. The practical implications of program emphasis on involving particular owner/operator firms include expanding the range of program instruments required, since some program incentives (such as loan guarantees) are more useful to some potential firms than others.

3. Observation drawn from interviews during February through May of 1980 with AE&C personnel at Badger America, Inc., Flour Corporation, and Stearns-Rogers, Inc.
4. For example, different technologies for coal-based synfuels require different amounts of engineering effort. See Bechtel (1979) and Dinneen, Merrow, and Mooz (1981).
5. See, for example, the Cameron Engineers study (1979) or various Rand Corporation studies. The key role of mega-project management teams was consistently asserted in interviews with AE&C firms.
6. There is some evidence to suggest that the third model may be important. For example, during discussions with the author, management personnel at one AE&C firm suggested that the company would be much more willing and capable of taking on responsibility for a $3 to $4 billion full-scale, coal-based synthetic fuel project after the company had initial experience with some synthetic fuel project at a more modest scale, say $1 to $1.5 billion. Presently, the company has limited experience with projects at the $500 million scale and above.
7. A price of $5 per barrel over the thirty-year life cycle of a synfuels plant beginning operation in 1990 equals a net present value of $0.8 billion at a 5 percent discount rate. See Chapters 2 and 3 for details of the calculation.
8. These specialized constructions trades were highlighted in the Bechtel study for the Department of Energy, which examined a 1 MMBD deployment during the 1980s followed by an additional 2 MMBD in the 1990s. Supplies of these skills could be taxed even in the industrialized Ohio Valley region. Other construction trades noted as potential constraints included iron workers and carpenters. See Bechtel (1979: 4-21).
9. Observation made by M. L. Mosier, President of the National Constructors Association (NCA), during interview in April 1980.
10. Interview with M. L. Mosier, President of the National Constructors Association, in April 1980.
11. Discussion with personnel at Fluor Engineers and Constructors, Irvine, California, April 30, 1980.

12. The Bechtel study of deployment requirements does not distinguish among regions for most equipment and materials, despite the region specific capabilities of their Energy Supply Planning Model. Bechtel notes that material and equipment "items are generally transported nationally and their use is not restricted to their area of production" (Bechtel 1979: 4-26).
13. Many of these demand and supply variations that complicate the assessment of shortages and constraints are well illustrated in the Bechtel study of impediments to synthetic liquids production. For example, adequacy of pressure vessel supplies depends on nuclear industry demand, use of multiple shifts in existing pressure vessel plants, subcontracting of some job components, and international procurement. See Bechtel (1979: 4-36–4-48).
14. Even though most hardware inputs are short lead-time concerns, development of infrastructure should yield some benefits by reducing factor cost inflation. Although supplies of short lead-time inputs can be quickly expanded relative to the time the inputs are needed for a project, such expansion probably will be associated with cost increases as suppliers attempt to quickly recoup the capital costs of capacity expansion via price increases. Labor productivity or product quality also may lag, due to new workers, multiple shifts, and so forth. Thus, as for other inputs, material and component constraints are most likely to translate into cost increases for synthetic fuels production rather than absolute limits on the number of plants possible. Enhanced deployment capabilities should increase the size of the production increments possible before such cost increases are encountered, and reduce the magnitude of the cost increases that may occur for any given deployment rate.
15. Such short lead-time concerns do require careful planning and attention during implementation of either an initial production program or a surge deployment effort. For example, valve manufacturing capacity can be increased, but clear, credible incentives and signals must be given to suppliers if they are to expand capacity to meet the requirements of major deployment efforts.

10 APPLYING THIS ANALYSIS TO THE U.S. SYNTHETIC FUELS CORPORATION PROGRAM

Much of this discussion has been purposely generic. The analysis has examined the general benefits, costs, and issues for initial synthetic fuel production efforts without special attention to particular programs or proposals for obtaining that experience. The observations and conclusions developed in this study could be applied to almost any program to promote, in advance of clear economic and market competitiveness, initial production of synthetic fuels. The basic framework also could be applied to major energy options other than coal-based synthetic fuels.

This chapter moves away from this generic orientation and develops some of the implications of this analysis for its most immediate and important implementation context—the program for initial synthetic fuels production authorized by the Energy Security Act of 1980 (ESA). Basic elements of the ESA program are described first, followed by an outline of the implications of this analysis in four major areas.

THE ENERGY SECURITY ACT PROGRAM

Basic Elements of the Program

The Energy Security Act of 1980 (Title I, Part B) establishes the United States Synthetic Fuels Corporation (SFC) as a "special purpose Federal entity to carry out the national synthetic fuel development program" outlined in the act (Conference Report 1980: 203). The legislation provides for a nominally two-phase effort oriented toward achieving goals for "synthetic fuel production capability of 500,000 barrels per day of crude oil equivalent by 1987 and 2,000,000 barrels per day by 1992" (Conference Report 1980: 208). The definition of "synthetic fuel" established in the act encompasses many resource bases and synfuel products in addition to the coal-based synthetic fuels focused on in this study. Thus, the estimates of costs, benefits, and appropriate program size apply primarily to a subset of the total range of synfuel resource bases included in the SFC program.[1]

An important milestone between the first and second phases of the initial production program is the preparation of a "comprehensive production strategy" for achieving production goals. The SFC is required to prepare and submit this strategy to Congress for approval by July 1984.[2] Acceptance of the comprehensive strategy is to be accompanied by additional appropriations as necessary to implement second-phase efforts according to the approved strategy.

The comprehensive strategy is to (1) set forth the "recommendations" of the SFC's board of directors on the corporation's "objectives and schedules for their achievement";[3] (2) provide a number of reports based on experience with first-phase projects on economic and technical performance as well as environmental effects; and (3) provide "recommendations concerning the specific mix of technologies and resource types to be supported during the second and subsequent phases" of the SFC efforts (Conference Report 1980: 209). Thus, the comprehensive strategy and its approval process presents an opportunity for the SFC and Congress to reappraise and redirect the U.S. program for initial synfuels production.[4]

Tensions Among Multiple Purposes

A review of the Energy Security Act reveals unresolved tensions among several potentially competing program purposes. The Act leaves important choices regarding program design and implementation emphasis to SFC discretion. The mandate for "commercial" production projects that support substantial production levels or "goals" is clear. However, other program purposes, clearly recognized in the legislation and related legislative history, may compete with a production emphasis.

For example, although references to production goals are repeated at several points, the act also clearly states that first-phase efforts should go beyond simple production and give attention to establishing the information and experience basis for improved and expanded production efforts in a second phase (Conference Report 1980: 209). A learning function for first-phase efforts is evidenced by repeated emphasis on using a diverse range of technologies and resource bases.[5] Recognition of the need to build improved deployment capability is indicated by the suggestion that projects should contribute toward development of "an industry infrastructure to support achievement of production goals" (Conference Report 1980: 211). Moreover, the reappraisal process represented by the "comprehensive production strategy" and its approval process reinforces the information and learning purposes of first-phase production efforts. The strategy process encourages the SFC to define appropriate goals and approaches for carrying out its mission.

RELATING THIS ANALYSIS TO THE SFC PROGRAM

Important benefits from an investment help define the purposes and objectives for an investment program. Clear articulation of purposes may help identify ways of increasing benefits or avoiding unnecessary or unproductive costs. Part I of this study describes important benefits—and, hence, important purposes and objectives—for the SFC program. Part II develops an appropriate analytical tool for assessing how large an investment program is appropriate given the many uncertainties and the merits of emphasis on particular sources of strategic insurance benefits. Part III develops issues for doing better or improving the cost-effectiveness of whatever size investment program is undertaken. This section outlines the implications of this study for five major aspects of the SFC program: program purposes, production goals and timing, comprehensive production strategy, program design and implementation provisions to enhance benefits, and SFC organizational capabilities.

Program Purposes

Definition of program purposes and the relative emphasis among competing objectives is an important step in program implementation. The strategic insurance benefits of "learning" and "deployability" developed in this study articulate further the concern for "technological diversity," a phased program structure, and development of an "industry infrastructure" that the Energy Security Act notes, in addition to production goals, as key concerns for program implementation.

The "strategic capabilities as insurance" view of program purposes provides rationale that apply specifically to the major long lead-time U.S. energy option represented by synthetic fuels from coal. Even if synfuel production costs—and resulting price guarantee or subsidy levels—exceed most plausible expectations for the long-run oil price trajectory and the oil import premium, the strategic insurance role shows why and how at least some practical production experience is usually a good idea. Given the uncertain and volatile world we face, one need not believe that synfuels are a desirable or inevitable alternative to recognize the value of at least a little initial practical experience. Such a prudent insurance capability purpose provides a more stable basis for the long-term commitments required of both the public and the private sector in the face of the "crisis" and "glut" history of both the world oil market and of U.S. energy policy.

Conflicts and tradeoffs among various program purposes are illustrated by the phases of the SFC program. For example, the first-phase production goal of 500,000 barrels per day probably exceeds the level of production and experience required to realize most of the learning opportunities available from an initial group of plants. The legislative history suggests a first-phase emphasis on quick learning with "technological diversity" and a special "fast start" program (Conference Report 1980: 188, 213). Such an emphasis on learning is more effectively served by a larger number of relatively small commercial plants rather than fewer full-size plants that emphasize production volume. However, learning-oriented, smaller projects provide contribute less substantially or certainly to the attainment of ambitious production goals.

The contrast between a strategic learning purpose and a production volume purpose is more striking for the second-phase production increment of the SFC program (1.5 million barrels per day additional production between 1987 and 1992). Achievement of the second-phase production goal would require the construction of the equivalent of thirty full-size plants between 1987 and 1992. If such a large and intensive deployment effort relied primarily on the coal resource base, it would involve many projects using substantially the same technology. The incremental learning benefits from such simultaneous or parallel projects would be small.

If we are fortunate and oil prices remain stable or decline, coal-based synfuels production may not yet be economic by the late 1980s or mid-1990s. Large production volumes—such as those envisioned by the second phase of the SFC program—might require production subsidies that would not be warranted by the benefits of reduced oil imports alone. A continuing production program emphasizing learning as a strategic insurance investment would recognize the sequential nature of the learning process and might substitute a relatively small third (or even fourth) sequential production phase in lieu of the large second-phase production expansions provided in the Energy Security Act. Such a continuing insurance capability program might avoid the atrophy of AE&C and other people-based capabilities that are key long lead-time constraints for synfuels option.

However, if synfuel costs and oil price relationships have made commercial production economic without subsidies, production should expand without any special government incentive program. Alternatively, if coal-based synfuel costs are so high as to make a surge deployment effort very unlikely, the value of second-phase or continuing investments in insurance capabilities with the coal-based synfuel should be critically assessed.

The deployability purpose introduces additional issues and tradeoffs. The ambitious production levels sought in the Energy Security Act program may provide substantial infrastructure development. If development of experienced AE&C capabilities is a key deployment constraint, however, the production levels sought in the first and second phases of the Energy Security Act may prove overly ambitious. Moreover, development of infrastructure may not be fostered as much by high production levels as by the involvement of key types of firms in appropriately sized and structured projects.

Neither the legislation nor this analysis resolves these tensions and tradeoffs for program purposes, design, and implementation. Ultimately, the tradeoffs and implementation provisions are the province of SFC decisionmakers (the SFC Board of Directors) and those who oversee the SFC program.

Fortunately, despite the stylized categorization that helps clarify distinct purposes, many program design issues or project selections serve multiple purposes and need not raise either/or choices. For example, relatively small initial projects could simultaneously support diverse learning as well as expand the number or capacity of AE&C firms with substantial experience and capabilities for future deployment.[6] Moreover, a portfolio approach to program design makes uniform emphasis on a particular source of benefits unnecessary and probably undesirable. Such a portfolio approach would incorporate some full-scale projects oriented toward production volume and some projects serving learning or infrastructure development purposes. The framework developed in this study helps define important attributes of the SFC portfolio. Different project types and purposes might require different incentive structures and contractual arrangements.

Phased Production Goals and Timing

Size. Part II of this study develops a decision analysis approach for assessing "How large?" an initial production program is appropriate. The systematic analysis of expected benefits and costs across many scenarios and views of key uncertain parameters and probabilities suggests that a moderate-scale initial production program may be better than either a small program that provides basic information and initial learning, or a large program that reaches one million barrels per day production capacity by 1990.

These decision analysis results apply to the first phase of an initial production effort that takes the entire 1980s decade for design, construction, and initial operation. A production expansion decision for a second phase occurs in 1990. The analysis results have strikingly different implications for the production levels defined as Phase I and Phase II goals in the Energy Security Act. If the first phase of the SFC program (1980–87) is viewed as the first decade (1980–90) assessed in the decision analysis, then the size of the first phase provided in the act—500,000 barrels per day—agrees with the moderate program size, which is a robust choice in the formal decision analysis results.

However, the Energy Security Act's production goal of two million barrels per day by 1992 at the end of the second phase is two times larger than large program of one million barrels day capacity by 1990 examined in Part II's decision analysis.[7] Chapter 3 showed that the levels of factor cost inflation for such a large program are likely to be substantial. The decision analysis showed that the added costs of factor price inflation generally more than offset any added benefits from production and infrastructure development. A large program—such as that envisioned by firm commitment now to the second-phase production goals—is likely to be worse than more modest efforts or no initial synfuels production in the 1980s at all.

Timing. The decision analysis results recommend a moderate initial production goal. Assessment of the design of the Energy Security Act program varies with the timing of production phases. The size of the first phase may be about right. However, the second-phase production goal in 1992 constitutes an overly accelerated schedule that is, at best, difficult to achieve. If first-phase plants take five to seven years to reach initial operation, second-phase plants must closely overlap with first-phase projects to reach the 1992 target date. While such a tight schedule may be achievable, it will certainly require expeditious progress with first-phase plants as well as innovative planning for second-phase projects. Second-phase project selections and design decisions will have to be made very soon after (or possibly before) start-up of first-phase plants.

Such a tight schedule is unlikely to reap many of the opportunities for sequential learning from first-phase plants. Such learning should improve the second phase in three important ways: (1) by clarifying appropriate program size as much better information on synfuel production costs becomes available; (2) by improving technology selection ("selection learning"); and (3) by fostering better versions of technologies ("technology improvement learning").

The production goal sought for the second phase of the ESA program may or may not be appropriate, depending on synfuel costs and oil market developments. However, given the uncertainties and the advantages of a phased decision and program structure, commitment now to such an ambitious 1992 goal for second-phase production expansion is premature.

Comprehensive Production Strategy

Timing. Nominally, process for the comprehensive production strategy presents an important opportunity for reassessment and clarification of production goals, emphasis among purposes, and other issues for the SFC program. However, the timing prescribed by the legislation for the strategy is even more problematic than the schedule for realizing second-phase production goals.

The practical operating experience required to report on technical and economic feasibility, environmental performance, recommended production schedules, and technology/resource portfolios will not be available by the mid-1984 date scheduled for submission of the strategy. The Energy Security Act recognizes the importance of a strategy based on sound information and provides for an extension of time for the comprehensive strategy if the SFC "determines that an adequate basis of knowledge has not yet been developed upon which to formulate and implement a comprehensive strategy."[8] However, the maximum extension allowed—one year or from mid-1984 to mid-1985—is very unlikely to provide sufficient time to acquire the practical operating experience necessary to develop a strategy incorporating the "adequate basis of knowledge" envisioned by the legislation.[9] Any strategy developed before practical operating experience is available may have to be revised if the actual performance of various technologies is different than that assumed in the strategy. A strategy that takes advantage of opportunities for sequential decisionmaking and can incorporate the improved information and operating experience from initial projects could not be completed until approximately seven years after program initiation.

Content of the Comprehensive Strategy. Irrespective of timing, the concepts, approach, and analysis developed in this study can contribute to the formulation of a sound production strategy. The decision analysis approach provides a ready analytical framework for further development and articulation of a phased production strategy that changes in response to key information and outcomes. The decision analysis model provides an analytical tool that can reflect both key uncertainties and a flexible strategy for expanding production levels.

The comprehensive production strategy should be more than a scheme for reaching established production goals. It can incorporate evolving information concerning major uncertainties that affect appropriate production levels and program emphases. For example, if first-phase experience shows synthetic fuels to have moderate costs or shows the advantages of a particular technology, then the production strategy for the second phase can be appropriately more aggressive and emphasize better technologies. On the other hand, if the basic cost level for synfuels is very high or if a particular synfuel technology is relatively high-cost, the comprehensive production strategy should be appropriately modified in response to such information.

There may be particular "signpost outcomes" that change preferred strategies and should be reflected in a dynamic strategy. For example, Part II shows that the decision analysis model can be used as a tool to assess the levels of factor cost inflation required to shift preferences for program size. If a production objective appears likely to exceed the levels of acceptable factor cost inflation, a more moderate objective may be appropriate. The decision analysis tool could be applied to other issues which have strong implications for appropriate choices.

Program Design and Implementation Provisions for "Doing Better"

This study has shown that clear articulation and assessment of important program benefits and costs can suggest approaches for improving the cost-effectiveness of any initial synfuels production effort. Attention to each major source of benefits raises issues for program design to enhance expected benefits.

Selection Learning. Attention to selection learning—knowing which technologies or approaches to employ in future deployment efforts—suggests the need for provisions to assure that appropriate information from initial production efforts is developed and made available to the unknown future decisionmakers whose choices may benefit from that information. For example, information coming from an initial production project should be structured to help future decisionmakers know whether a high cost outcome was due to poor technology or poor project execution. Without some attention to information outputs now, it may be difficult to assess some years in the future why a particular project cost 30 percent more than originally budgeted and took six months longer. Was the increase in cost and time caused by bottlenecks, or expensive labor, or bad management, or poor technology, or some other factor?

Technology Improvement Learning. Attention to technology improvement learning highlights the value of appropriately phased projects where opportunities for sequential learning are exercised. Given limited resources for insurance investments in learning with the synfuels option, sequential projects are likely to be more cost-effective than nearly identical plants built simultaneously in order to reach arbitrary production goals quickly.

Also, technology improvement learning is useful primarily if it remains relevant to future deployment efforts. Where new technologies are likely to dominate future deployment efforts, insurance investments in commercial production experience may remain most relevant if they keep up with the evolving technology frontier. For example, direct liquefaction technologies (SRC, H-Coal, Exxon Donor Solvent) represent a fundamentally different approach that is not yet commercially proven. If a major technology option appears to be sufficiently economically attractive to draw the substantial equity co-funding from the private sector required by the Energy Security Act, the SFC may wish to consider funding appropriately structured projects with technologies that have higher technical risks. Thus, it may be appropriate for some initial production projects to be "technologically aggressive" in order to provide practical experience with technology that is likely to be important to future deployment efforts.

Technology improvement learning may also be fostered by project selection and program structure which gives attention to the incentives of project participants. For example, some firms may have an entrepreneurial interest and capability for advancing technology, while the limits of other firms' interests may extend only to the initial project itself. Different forms of economic incentives, financial assistance, or contractual provisions may be appropriate for commercial-scale projects emphasizing or facilitating technology innovation.

Infrastructure Benefits and Factor Cost Inflation. Concern for deployability effects focuses program design and implementation efforts on linkages between initial production experience and future deployment capability. Some projects or program design provisions—especially those related to developing additional actors with the capability to develop, design, and execute a synthetic fuels project—could be relatively effective for relaxing potential deployment constraints.

For example, if the level of initial synfuels production effort is limited by budget or other constraints, we may want to assure that a subset of the initial production projects provide experience to AE&C firms which, while nearly qualified, have not yet undertaken projects having the scale or complexity of a synfuels plant. The initial production projects might train these AE&C firms so that they become fully capable of undertaking full-size synfuels projects in the future. Such emphasis on expanded deployment capabilities could be implemented via appropriate provisions for project selection and/or incentive contracts.

Characterization of existing deployment constraints can illuminate the size of an initial production effort for which factor cost inflation is likely to become a serious concern. In particular, the analysis in Chapters 3 and 9 suggests that the existing capacity of AE&C firms that are well-qualified to undertake efficiently a major coal-based synfuels plant may allow about ten projects. While it may be possible to leverage existing capability (and some strain on the infrastructure may be necessary to expand capabilities for the future), initial production efforts that substantially exceed such a long lead-time constraint are likely to incur inefficiencies and substantially increased costs.

The general program design concepts and provisions noted above illustrate only a few major issues for cost-effective program design. Part of the strategic analysis and program design problem for the Synthetic Fuels Corporation involves further assessing these issues, arriving at policy judgments, and designing implementation provisions.

SFC Organizational Capabilities

The purposes and orientations for an initial synthetic fuel production program also have implications for organizational capabilities required for the Synthetic Fuels Corporation. Since initial synthetic fuel production may not be clearly economic to potential private investors, a large part of the SFC function is the delivery of necessary economic and financial incentives to private project developers. The organizational capabilities required for this function include those usually associated with banking and investment banking as well as the legal skills required for major project contracting between two organizations. These functions may well dominate the SFC's limited staff allotments.

However, the additional public objectives for the SFC program create the need for effective staff capabilities to assist in characterizing management options and planning the SFC program. These planning and policy functions for the portfolio of SFC programs and related social purposes are outside the normal domain of straightforward banking or contracting operations. Some of the analysis required is cross-project or cross-industry policy analysis related to *program* benefits, costs, and purposes rather than to evaluation of the technical or financial merits of an individual *project*. Organizational capabilities to perform this strategic planning, program evaluation, and public policy function should receive appropriate emphasis.

In addition to policy and planning capabilities, attention to learning benefits may heighten the need for technical evaluation capabilities (or perhaps, liaison with organizations having those capabilities). For example, assessment of the commercial readiness of a promising, but unproven, technology may well require technical and engineering capabilities beyond those commonly needed for a banking function. Effective implementation of program design approaches for improving cost-effectiveness may require carefully developed solicitation, evaluation, and incentive contract negotiation provisions as well as related staff capabilities that can be responsive to broad program objectives.

In general, the framework for program benefits and purposes developed in this study implies important roles for organizational capabilities beyond those required for normal banking and contracting functions.

CONCLUSIONS FOR THE ENERGY SECURITY ACT PROGRAM

The Energy Security Act (ESA) represents a good starting framework for an initial synthetic fuels production effort. Reflecting the value of the intensive review and compromise inherent in the legislative process, the Energy Security Act improves upon the House, Senate, and Carter administration proposals that it melds together. In particular, the program's phased structure with its mid-stream reassessment and "comprehensive production strategy" recognizes the limited information upon which enormous investment decisions must be based. Initial production experience will reduce key uncertainties, clarify appropriate choices, and provide improved and expanded options.

Unlike some proposals for several million barrels per day of synfuels production capacity within a decade, the production goals in the ESA are plausibly achievable, although a firm commitment now to a second-phase goal may be premature. This study's cost-benefit assessment suggests that the ESA's first-phase production goal is likely to yield benefits that exceed costs. Although smaller initial efforts often yield similar levels of net benefits, larger programs are likely to incur great additional costs for few added benefits.

Timing is the primary problem for the ESA program. The timing between production phases is overly tight; a program meeting the legislated schedule is likely to forego substantial opportunities for sequential learning. The statutory timing for the comprehensive production strategy is particularly incongruous with the technological realities for the time required to design, construct, and operate a synfuels plant. The comprehensive production strategy should be based on practical operating experience from several first-phase plants. This experience will not be available until approximately 1987. Firm, credible commitments to first-phase projects are necessary, but a premature comprehensive strategy provides a poor basis for the major long-term decisions and commitments that Congress, the administration, and the private sector must make for any second-phase efforts.

The Energy Security Act also provides only limited guidance on the tensions and tradeoffs among the various program purposes it recognizes. The act wisely leaves many important choices and substantial discretion to the Synthetic Fuels Corporation, but its apparent emphasis on production goals and production volume as a measure of program accomplishment and value could encourage the SFC to discount other important strategic benefits, such as learning and infrastructure development. Those purposes should weigh significantly in SFC choices.

Cost-effective program design requires that the SFC develop a well-grounded, carefully considered statement of mission. Emphasis among competing program purposes may require clarification. Ultimately responsible decisionmakers—such as the SFC board of directors—must make tradeoffs. The perspective and tools developed in this study can aid in articulating and evaluating the tradeoffs and in developing a program that increases program benefits or avoids unwarranted costs.

NOTES

1. See Conference Report (1980:201) for the definition of "synthetic fuel" provided in the act.
2. The comprehensive strategy is required within four years of the date of enactment (June 30, 1980). A one-year extension is allowed if sufficient information is not available to develop the strategy. Thus, the latest allowable date that meets the statute's prescribed timing is July 1985. See Conference Report (1980:209).
3. See text of Energy Security Act (1980), Title I, Part B (The Synthetic Fuels Corporation Act of 1980), Section 126.b. (3), or Conference Report (1980: 209).
4. Actually, the approval process for the comprehensive production strategy explicitly seeks to avoid modifying the Energy Security Act itself. The strategy is to be accepted (or rejected) as a whole by joint resolution of Congress. Modifications are allowed primarily for the total authorization level. The process for approving the joint resolution explicitly prohibits amendments that limit the use of the additional budget authorizations to particular uses or projects.
5. For example, see references to diversity in Title I, Sections 126 and 127, of the Synthetic Fuels Corporation Act (1980) or references in the Conference Report (1980:209, 210).
6. As well as provide developed sites where relatively rapid "brownsite" expansion of an existing facility is possible.
7. However, note that the decision analysis results apply to relatively high-cost coal based synfuels, while the definition of "synthetic fuels" in the Energy Security Act includes additional resource bases such as heavy oil and oil shale.
8. Energy Security Act, Title I, Part B (The Synthetic Fuels Corporation Act of 1980), Section 126.d.(1).
9. The requirement for a production strategy within four to five years of enactment is clearly problematic if the strategy is to be based on operating experi-

ence (as distinct from design and construction experience) with first-phase projects. Section 126.d.(1). does not make clear whether an additional one-year extension can be granted after the initial maximum one-year extension has been allowed. Of course, if multiple year-by-year extensions can be granted, the comprehensive production strategy can be deferred legally until the 1987 timeframe, when initial production experience from first-phase projects is likely to be available.

According to Dr. E. H. Blum, a former Department of Energy official who was integrally involved in the development of the legislation, the timing for the production strategy requirement reflected both a congressional intent to press for expeditious program progress and a compromise between the House, which wanted a second-phase strategy within three years, and the Senate, which maintained that five years was appropriate timing for second-phase strategy and commitment.

11 CONCLUSIONS

A RATIONALE AND STRATEGY FOR GETTING STARTED

Initial commercial-scale production of synthetic fuels requires enormous investments of the nation's limited resources. Such investments should not be made without careful consideration of potential benefits and costs. The dynamic and uncertain world we face makes the benefits and costs of synfuels production highly uncertain. Despite these uncertainties, or perhaps because of them, the United States is preparing to make the major investments necessary to start improving experience, understanding, and capability with the major energy option represented by coal-based synthetic fuels. Given the magnitude of the investments and uncertainties, it is important to define a clear rationale and strategy for getting started.

Each part of this study has described major elements, analytical approaches, and issues for such a strategy. Part I develops a framework for potential benefits and costs of initial synthetic fuels production experience. "Production benefits"—the value of each barrel of energy produced—are uncertain and provide no rationale for special emphasis on synfuels. "Information benefits" are widely recognized but require only small initial production efforts. The key source of benefits underlying this study are "surge deployment benefits"—improved national capabilities to build better and more synfuel plants should unhappy events force aggressive synthetic fuels deployment in the future. Because of the long lead-time for learning relative to the potential for rapid change in energy markets, surge deployment benefits provide special rationale for investments in initial synfuels production experience. Moreover, surge deployment benefits help define a framework for assessing "How large?" a program is appropriate and "How to do better?"

The "strategic capabilities as insurance" framework for this study explicitly recognizes key uncertainties and the potential dynamics of the oil market. Likely subsidies for initial coal-based synfuels projects are the insurance "premiums." Improved surge deployment capabilities provide the insurance "payoffs" if some unfortunate events occur. Answers to the question of how much insurance to purchase depend on (1) the relationship of payoffs to the level of initial production experience; (2) the cost or premiums required to obtain that experience; and (3) the likelihood that an insurance event will occur. The majority of "learning" payoffs accrue from a small program, but the infrastructure development (or "deployability") payoffs may be much greater for larger initial production efforts. The cost for the insurance premium for production subsidies increases greatly as large production efforts induce "factor cost inflation" by over-straining the existing capabilities of the industrial infrastructure.

Part II of this study develops and applies a decision analysis approach and model for assessing this complex insurance problem. The model systematically assesses many combinations of key uncertain outcomes, such as synfuel production costs, long-run oil price trends, and the possibility of rapid increases in the market price or security costs of imported oil. Phased decisions can respond to evolving information about key uncertainties, to the occurrence of an oil price jump, or to other insurance events. Results necessarily depend on input parameters and probabilities. The decision analysis model is primarily a tool for assessing the implications of alternative views of key uncertainties and our energy future. However, sensitivity analyses detailed in Part II suggest that a moderate scale initial production effort is likely to be a robust choice across many expectations for insurance payoffs, premiums, and events.

Technically, a "moderate" program is defined as 500,000 barrels per day of coal-based synfuels by about 1990. However, the analysis shows that production volume is not the primary source of benefits that make one program size better than another. Rather, the advantage of a moderate initial production effort stems from achieving substantial learning as well as some infrastructure development without incurring large near-term costs from factor cost inflation. The "moderate" program size recommended by the decision analysis assessment represents not a particular production volume as much as a program that (1) realizes much of the learning possible in an initial phase; (2) achieves modest infrastructure development; and (3) incurs only acceptable levels of factor cost inflation (less than 10 to 20 percent).

These quantitative assessments of "How much?" insurance is appropriate are illuminating. However, the usefulness of the "strategic capabilities as insurance" concept for the benefits and purposes of initial synthetic fuel production efforts extends beyond such quantitative results to suggest qualitative program design provisions for doing better.

Part III of this study shows that careful assessment of the linkages between initial synfuels production experience and improved surge deployment capabilities can provide useful guidance for program design. Analytical questions and program implementation provisions for achieving more payoffs with a given insurance premium are described.

The approach, analysis, and exposition set forth in this study have been consciously conceptual. Rather than focusing on a particular proposal or program, it has sought to articulate a well-founded general framework and approach for assessing the benefits and costs for the nation of initial practical experience with synthetic fuels production. This study does not automatically assume that a government program is necessary or desirable. Rather, it seeks to develop an appropriate framework for assessing the value of initial synfuels production efforts and for improving the design of any programs that might be undertaken. One important caveat must be noted before outlining eight major lessons from this analysis for a sound strategy for getting started with synfuels.

NATIONAL PURPOSES NEED NOT IMPLY PUBLIC MANAGEMENT

Strategic benefits to the nation as a whole could justify a special program to promote initial synthetic fuel production activities. This study has identified approaches for improving the cost-effectiveness of an initial production effort via program design and implementation provisions that accent key national benefits and purposes. Some of these program design provisions may seem to imply detailed government or SFC planning and management of initial production projects. However, program design and implementation attention to social benefits and purposes need not lead to government or SFC involvement in the detailed development and implementation of initial production projects.

We should recognize the significant efficiencies of market processes and unfettered private decisionmaking for providing excellent assessments of which processes and projects should be included in any initial production efforts. Moreover, the discipline of the marketplace and the exposure of substantial amounts of private money to risk should help maintain sound project development and execution incentives. Accordingly, there may be little need for detailed government review or oversight.

Also, government management prerogatives may be unwarranted and undesirable. Many of the benefits of initial synthetic fuel production efforts stem from undistorted experience with the major technical and organizational issues involved with the basic activity of commercial synthetic fuel production. Detailed government involvement in the development or execution of projects may well distort the experience and frustrate the realization of some program benefits and objectives.

More importantly, however, many of the program design and implementation provisions enhancing social benefits and objectives can be defined in broad terms which need not direct detailed project development and implementation choices. The government or SFC has a planning interest only in the structure and attributes of its broad *program*, not in the details of particular *projects*. For example, the programatic attribute of resource base diversity can be defined by provisions requiring some projects in each resource base. Technology diversity, while more difficult to define than resource bases, can also be demarked with rather gross, objective distinctions, such as "coal to liquids" or "coal to liquids via indirect liquefaction." Accordingly, detailed evaluation and selection among technologies may be neither necessary nor appropriate.

The government's primary interest is in promoting experience that leads to improved national capability with the synfuels option.

The capabilities are and should remain located in the nation's private sector. Moreover, it is likely that the projects proposed by private firms will result in a range of program attributes, such as technology diversity, that acceptably serve the national interest.

However, even a public interest or objective that translates into a constraint on private choices need not imply detailed government involvement. For example, a social interest in obtaining experience with a particular technology need not lead to a government involvement in detailed design or management of the project utilizing that technology. Similarly, the public interest in an objective such as enhanced deployability could be implemented by program design provisions that encourage involvement of new actors—such as major architect/engineering firms that are not yet fully "synfuel-capable"— without specifying which of a large number of possible AE&C firms is utilized by the private sector project sponsor.

Moreover, some social objectives can be advanced without constraining private choices. Important social purposes can be served simply by providing information on the experience resulting from unfettered project execution. Future investment decisions may be improved if relevant information can be made available to unknown future decisionmakers. For example, high construction labor rates or bottlenecks causing delays in construction schedules could cause a technology to appear to be more costly than it would be under other circumstances. Information that allows future decisionmakers to distinguish project execution difficulties from the costs inherent to a technology could be valuable. Such information is not highly proprietary or difficult to obtain if relevant information needs are well-defined at the outset of project efforts and appropriate arrangements made. For many types of information, elaborate efforts to gather or verify information are neither necessary nor appropriate. Where the acquisition or verification of information requires major costs or treads on important proprietary interests, government should objectively assess whether the information is worth the effort and potential distortion necessary to obtain it.

Most, if not all, of the overall social benefits of initial synthetic fuel production experience can be obtained without elaborate program management or implementation provisions. Certainly, attention to public benefits and purposes need not imply any government or SFC profile in the design or execution of individual projects. The government planning interest is primarily in the program-wide portfolio and the availability of information from the experience.

Moreover, the substantially generic character of the basic activity of synthetic fuel processing allows many purposes to be served simultaneously by an individual project. The first few projects in an initial production program will serve many attributes of a strategically designed portfolio for initial production experience. Accordingly, it may be both possible and appropriate to defer decisions on portfolio emphasis subject to the development of both the analytical support for choices and a larger variety of projects among which choice is necessary. Only a few well-developed projects are immediately available to start the first phase of the SFC program. Thus, some of the analysis and lessons developed in this study may be more relevant to the later part of the first phase or to the second phase of the SFC program.

A final caveat must reiterate the concept of cost-effectiveness for particular program design provisions. Many of the approaches for enhancing the benefits of initial production experience involve minor adjustments in project selection or implementation. However, the tendency to overplan should be resisted. Real but potentially minor planning advantages should not carry inappropriate weight relative to unfettered competitive market processes and private decisionmaking in the development and implementation of initial production projects. The test of cost-effectiveness should be applied to program design and implementation provisions. The added costs of particular provisions should be compared to their expected incremental benefits.

LESSONS FOR A STRATEGY FOR GETTING STARTED

Rather than attempting conclusive assessments of issues, this study has emphasized developing a framework and approach for assessing benefits, costs, and program design issues. The framework for the benefits and purposes of initial synthetic fuels production efforts is as important as the quantitative and qualitative results that derive from that framework. It is important to have a clear sense of what we are trying to accomplish with initial synthetic fuel production efforts. It is important to get the rationale and rhetoric correct at the outset. Once a sound, well-accepted set of program purposes is established, many questions, answers, and strategies relevant to doing better become readily apparent.

The results of this study could be provocative. They comment on the merits of continuing the Energy Security Act program. They bear on appropriate goals and objectives for the multi-billion dollar investments envisioned by the SFC program. This analysis also has implications for SFC program design and the evaluation of individual projects.

Review and application of this analysis will certainly refine, extend, and improve each of its three major parts. The decision analysis tool for systematically reflecting key uncertainties and competing benefit-cost relationships could be further developed and applied to alternative program strategies rather than to only the "How large?" issue addressed in Part II. Phased or sequential strategies that reflect implications of information outcomes for key uncertainties can be described and evaluated. Priority elements of the analytical agenda for cost-effective program design can be assessed. In general, the overall framework, tools, and approaches outlined in this study can be further developed into practical program objectives, strategies, and implementation provisions.

These next steps are important. This study has merely outlined the major concepts, analytical approaches, and issues that are important elements of a sound rationale and strategy for getting started with synfuels—a synfuels *"start*egy".[1] Nevertheless, the conclusions of this analysis suggest a few robust principles or lessons for a synfuels *"start*egy".

Go, But Go Moderately

Both the words "go" and "moderately" are important. The decision analysis of insurance costs and benefits shows that some initial synfuels production experience is almost always better than no initial production in the 1980s. Only combinations of extremely high synfuel costs and low oil prices make the "no program" choice appropriate. However, the "go" part of the lesson should not be detached from the "moderately" part. The certain enormous potential of synthetic fuels seems to encourage overly ambitious production goals. If and when synfuels are clearly economic, ambitious deployment schedules may be appropriate. However, for insurance investments, we should not ignore the constraints that make doing more cost more. More production does buy more insurance—especially increased deployment capability. The added insurance payoffs occur in only some scenarios, however, while the added insurance premiums for large programs must be paid in all scenarios.

The decision analysis model looks systematically across many scenarios—some are favorable to synfuels, others unfavorable. The appropriate program size necessarily depends on expectations for key parameters and probabilities. Sensitivity experiments show that some initial production experience is almost always a good idea. A moderate program is noticeably better than a small program over a wide range of expectations. However, for many expectations and scenarios, a large initial production program yields substantially fewer net benefits that a more modest initial production effort and could be worse than no initial production in the 1980s at all. For large programs, the added costs of doing more generally outweigh the added insurance benefits. A small to moderate program is best for most views of the world.

Factor Cost Inflation Is a Critical Issue

"Factor cost inflation" is a summary word for the added costs of doing more at the same time. Large initial production efforts will increase the prices for the materials and services required to build synfuels projects. Bottlenecks and shortages lead to construction inefficiencies. Factor cost inflation rapidly increases the average cost and insurance premium for production subsidies for all plants built simultaneously during an initial production effort.

The decision analysis results demonstrate that factor cost inflation is a key indicator of appropriate program size. The added costs from factor cost inflation quickly offset potential future benefits from expanded infrastructure for surge deployment scenarios. Modest levels of factor cost inflation quickly make a moderate program preferable to a large program and, eventually, a small program better than a moderate program. Optimum initial production program scale is reached shortly after the additional costs due to factor cost inflation begins exceeding 10 to 20 percent.

The production volumes at which such levels of factor cost inflation would be encountered are uncertain. Since factor cost inflation is a key indicator of appropriate program size, further analysis and monitoring of factor cost inflation effects should be priority concerns for program design and implementation.

Infrastructure Benefits Tradeoff with Factor Cost Inflation

Deployability effects reflect the infrastructure development concept that links present production experience to expanding future deployment capability. The major advantage of larger initial production efforts is expanded infrastructure development. Deployability effects tradeoff with factor cost inflation to increase the value of larger initial production efforts. If future deployment capability is largely independent of initial production experience or infrastructure, then a small initial production program provides adequate insurance capability for surge deployment efforts. Learning and information benefits can be obtained via a small or modest initial production effort. However, if there is a strong linkage between initial production experience and expanding surge deployment capabilities, it may become worthwhile to buy more insurance. Up to a point, it may be appropriate to accept modest amounts of factor cost inflation in order to build infrastructure for expanded future deployment capability.

Potential deployability effects are not well understood. However, existing deployment constraints are likely to cause factor cost inflation for initial programs. Primary attention for strategic investments in infrastructure development should focus on those deployment constraints that can be relaxed only via investments *prior* to mounting an all-out deployment effort. Constraints on qualified architect, engineer, and constructor (AE&C) services for project design and management act both as an existing constraint on the size of an initial production program and as a focus for efforts to build infrastructure for the future. Better understanding of the tradeoffs between infrastructure development and factor cost inflation is important to improved program design.

Use a Truly Phased Program Structure

Synfuels development is fraught with market, economic, and environmental uncertainties. These uncertainties should not deter synfuels production commitments. Rather, the uncertainties reinforce the value of a phased program structure. Some initial production experience is necessary to reduce uncertainties about synfuel production costs and environmental impacts. The initial production phase also provides a basis for improved and expanded second-phase efforts.

However, full realization of the advantages of a phased structure, requires that the program structure be truly phased. Much of the learning and information from initial synthetic fuels production experience is revealed only with plant start-up and initial operation. The demarcation point between an initial phase and final planning and design for a subsequent phase should not come before initial operating experience is acquired with first-phase plants.

This observation has two important implications. First, the schedules outlined in the Energy Security Act of 1980 are not congruent with the underlying technological practicalities. This is probably true for the two-phase production goals of 500,000 barrels per day by 1987 and 2,000,000 barrels per day by 1992. It is certainly true for the comprehensive production strategy for the second phase required by 1985. An accelerated schedule for design to operation of a representative synfuels project requires five to seven years. As a result, information from first-phase plants is unlikely to be available to guide either the comprehensive strategy or the design of second-phase plants until the mid to late 1980s. If the program of the Synthetic Fuels Corporation is to be truly phased, the legislated schedules may require adjustment.

Second, the primary way to reduce the hiatus period, during which we have limited practical information and experience with the synfuels option, is to proceed expeditiously with a few "fast-track" initial projects. These projects provide primarily basic information and initial technology improvement learning. The volume, cost, or market value of production is not nearly as important as timing. The sooner we acquire initial operating experience, the sooner we are past the hiatus period during which any surge deployment efforts would incur the mistakes and costs of being necessarily based on limited practical experience.

Collect Information

The substantial uncertainties confronting synfuels investments magnify the importance of the information output of initial production projects. Moreover, many of the information benefits could stem from applying the experience from initial production efforts to surge deployment decisions by unknown decisionmakers at some undetermined future time. Important information from initial production expereience can be lost if it is not recorded, translated, and maintained in useful form. Thus, it is important to develop a history that collects and organizes information from initial production efforts. An information strategy should be developed. This strategy should identify the types of information relevant to future decisions, define provisions for obtaining the information, and develop mechanisms to maintain information in a format that can be readily applied as necessary. Potential uses of the information include near-term decisions and strategies for second-phase production efforts (the "comprehensive production strategy") as well as improved technology selection and design for future surge deployment efforts. Provisions for obtaining information should not unduly distort the experience that the information is supposed to characterize. The information strategy should also recognize important proprietary interests of private sector participants in initial production projects.

Respond to Information

Decisions on synthetic fuel production need not be made once for all time. A phased approach allows a dynamic production strategy, which can respond to evolving information about key uncertainties. Phased production strategies can be expanded or contracted in response to information that clarifies the relative economics and market role for the synfuels option.

The costs of ignoring information can be substantial. For example, the decision analysis assessment in Part II of this study shows that costs of several tens of billions of dollars (net present value) may be incurred if production levels are not adjusted in response to information that clarifies appropriate choices.[2] Information is valuable only if we adjust our strategy in response to it. Stable commitments to initial production efforts are necessary, but inflexible production strategies probably forego substantial opportunities for improvements in the scale, timing, or technology mix for synthetic fuel production efforts.

Consider the Advantages of Smaller Projects

Smaller initial commercial projects foster as much or more information and technology improvement learning as full-size plants. An initial production program involving a larger number of smaller plants might also increase the deployment infrastructure more effectively than a program with fewer large projects. Also, using smaller plants increases diversity and reduces the risks of being technologically aggressive or innovative with a subset of projects. The merits of the common industrial practice of starting small with first-of-a-kind production facilities should be explicitly considered.

Emphasize Future Capabilities with the Synfuels Option, Not Near-Term Production Levels

Costs of benefits from the physical barrels of synfuel production are uncertain. Substantial net costs or savings relative to oil are possible. Improved capabilities to move efficiently and aggressively with the synfuels option when clearly warranted are a more important output of initial production efforts than barrels of production. The type and quality of the experience and information from initial production efforts may well be more important than production volume. Capabilities are improved via significant commercial production experience. However, higher production levels do not correlate one for one with improved capabilities with the synfuels option. Emphasis on synfuel capabilities rather than on production volumes should allow us to achieve a higher insurance payoff for a given insurance premium or adequate insurance protection with a reduced insurance investment.

NOTES

1. William Hogan first coined the word "*start*egy" as a summary word for the rationale and strategy for getting started developed in this study.
2. A large program is best if synfuels costs are uncertain over a given range. However, if we know that synfuel costs exceed $65 per barrel, a small program becomes the best choice. The net losses of proceeding with a large program, even when we are sure synfuels are expensive ($65 per barrel), exceed $40 billion discounted present value.

REFERENCES

Baumol, William J. 1972. *Economic Theory and Operations Analysis.* Englewood Cliffs, N.J.: Prentice-Hall, Inc.

Bechtel National, Inc. 1979. "Production of Synthetic Liquids from Coal: 1980–2000, A Preliminary Study of Potential Impediments." Draft final report prepared for the U.S. Department of Energy's Office of Fossil Fuel Processing under contract number ET-78-C-01-3127-M001.

Bodde, David L. 1976. "Riding the Experience Curve." *Technology Review* 78, no. 5 (March/April), reprint.

Booz, Allen & Hamilton. 1979. "Analysis of Economic Incentives to Stimulate a Synthetic Fuels Industry." In *Synthetic Fuels: Report by the Subcommittee on Synthetic Fuels of the Committee on the Budget—United States Senate,* pp. 55–148. Washington, D.C.: U.S. Government Printing Office.

Braun, C.F. 1976. "Factored Estimates for Western Coal Commercial Concepts." ERDA/AGA Contractor Report.

Brusher, William F. 1979. "Home Office Productivity." *Proceedings of Engineering Construction Contracting Meeting.*

Bupp, Irvin C., and Jean-Claude Derian. 1978. *Light Water.* New York: Basic Books, Inc.

Cameron Engineers. 1979. "Overview of Synthetic Fuels Potential to 1990." In *Synthetic Fuels: Report by the Subcommittee on Synthetic Fuels of the Committee on the Budget—United States Senate,* Washington, D.C.: U.S. Government Printing Office.

Committee for Economic Development. 1979. *Helping Insure Our Energy Future: A Program for Developing Synthetic Fuel Plants Now.* New York: Research and Policy Committee.

CONAES Modeling Resource Group. 1978. "Energy Modeling for an Uncertain Future." Supporting Paper 2, Study of Nuclear and Alternative Energy Systems (CONAES), National Research Council. Washington, D.C.: National Academy of Sciences.

Conference Report. 1980. "Joint Explanatory Statement of the Committee of Conference." Report No. 96-1104, 96th Congress, 2nd Session. Washington D.C.: U.S. Government Printing Office.

Deese, David, and Joseph Nye, Ed. 1980. *Energy and Security.* Cambridge, Mass.: Ballinger Publishing Co.

Department of Energy. 1978. *International Coal Technology Summary Document.* No. 061-000-00206-5. Washington, D.C.: U.S. Government Printing Office.

——— 1979. "Energy Security Corporation Briefing Book," Unpublished draft manuscript prepared for Office of Policy and Evaluation (Blum/Ingerman). Washington, D.C.

———. 1980a. *Draft Policy, Program, and Fiscal Guidance: FY 1982-1986.* Washington, D.C.

———. 1980b. *Policy and Fiscal Guidance Energy Projections.* Appendix D, DOE/PE-0018, Assistant Secretary for Policy and Evaluation, Office of Analytical Services. Washington, D.C.

———. 1980c. "The Energy Problem: Costs and Policy Options." Staff working paper, Office of Oil Policy, Policy and Evaluation. Washington, D.C.

———. 1980d. *Reducing U.S. Oil Vulnerability: Energy Policy for the 1980s.* Analytical report to the Secretary of Energy, Assistant Secretary for Policy and Evaluation. Washington, D.C.: U.S. Government Printing Office.

———. 1981a. "Water Supply and Demand in an Energy Supply Model." DOE/EV/10180-2. Springfield, Va.: National Technical Information Service.

———. 1981b. *Energy Projections Through the Year 2000: An Appendix to the Third National Energy Plan.* Office of Analytical Services, Office of Policy, Planning, and Analysis, DOE-PE/0078. Springfield, Va.: National Technical Information Service.

Dineen, P., Edward Merrow, and William Mooz. Forthcoming. *Analysis of the Architect-Engineering and Construction Industry.* Santa Monica, Calif.: The Rand Corporation. Unpublished draft manuscript cited with special permission.

Energy Information Administration. April 1981. *1980 Annual Report to Congress, Volume Two: Data.* Washington, D.C.

Energy Research and Development Administration (ERDA). 1976a. *Synthetic Liquid Fuels Development: Assessment of Critical Factors.* ERDA 76-129. Washington, D.C.: U.S. Government Printing Office.

Energy Research and Development Administration (ERDA). 1976b. *Factored Estimates for Western Coal Commercial Concepts.* Contract report. Washington, D.C.: U.S. Government Printing Office.

Energy Research and Development Administration (ERDA). 1977. *Programmatic Environmental Impact Statement on Alternative Fuels Demonstration Program.* Washington, D.C.: U.S. Government Printing Office.

Energy Security Act. 1980. Public Law No. 96-294. Text as presented in "Energy Security Act: Conference Report", Report No. 96-1104, 96th Congress, 2nd Session. Washington, D.C.: U.S. Government Printing Office.

Enos, John L. 1962. *Petroleum Progress and Profits.* Cambridge, Mass.: MIT Press.

Federal Energy Regulatory Commission. 1979. "Opinion and Order on Requests for Certificates Involving Proposed Coal Gasification Project." Opinion No. 69. Washington, D.C.

Flour Corporation. 1979. *A Flour Perspective on Synthetic Liquids: Their Potential and Problems.* Irvine, Calif.

Gallagher, J. Michael, M. Carasso; R. Barany; and R.G. Zimmerman. 1976. "Direct Requirements of Capital, Manpower, Materials, and Equipment for Selected Energy Futures." Report prepared for Energy Research and Development Administration under contract No. E(49-1)-3794, PAE/3794-1, Bechtel Corporation. Springfield, Va.: National Technical Information Service.

Harlan, J.K. 1981. "Toward A Synfuels 'Start'egy: Benefits, Costs, and Program Design Assessments for Initial Synthetic Fuel Production Efforts." Ph.D. dissertation, Harvard University.

Hirschmann, Winfred. 1964. "Profit from the Learning Curve." *Harvard Business Review* (January/February): 125.

Hoffman, F.S., and M. Kennedy. 1981. "World Oil Market Trends and U.S. Energy Policy." R-2573, Santa Monica, Calif.: The Rand Corporation.

Hogan, William W. 1979. "Dimensions of Energy Demand." Discussion paper E-79-02, Energy and Environmental Policy Center, Kennedy School of Government, Harvard University.

———. 1980. "Import Management and OIl Emergencies." In *Energy and Security*, edited by D. Deese and J. Nye, pp. 261-301. Cambridge, Mass.: Ballinger Publishing Co.

Hottel, H.C., and J.B. Howard. 1971. *New Energy Technology: Some Facts and Assessments*. Cambridge, Mass.: The MIT Press.

Hotelling, H. 1931. "The Economics of Exhaustible Resources." *Journal of Political Economy* 39, no. 2 (April): 137-175.

ICF Incorporated. 1979a. "Imperfect Competition in the International Energy Market: A Computerized Nash-Cournot Model." Report to the Office of Policy and Evaluation, Department of Energy, Washington, D.C.

———. 1979b. "Oil Import Reduction: An Analysis of Production and Conservation Alternatives." In *Synthetic Fuels: Report by the Subcommittee on Synthetic Fuels of the Committee on the Budget—United States Senate*. Washington, D.C.: U.S. Government Printing Office.

Jacoby, Henry D., M.A. Adelman; B.C. Ball, Jr.; L.C. Cox; T.L. Neff; D.C. White; and D.O. Wood. 1979. "Energy Policy and the Oil Problems: A Review of Current Issues." Energy Laboratory Working Paper, No. MIT-EL 79-046 WP, Massachusetts Institute of Technology.

Joskow, Paul L., and Robert S. Pindyck. 1979. "Should the Government Subsidize Nonconventional Energy Supplies?" *Regulation* 3 (September/October): 18-24.

Joskow, Paul L., and George A. Rozanski. 1979. "The Effects of Learning by Doing on Nuclear Plant Operating Reliability." *The Review of Economics and Statistics* 61, no. 2 (May): 161-168.

Mansfield, Edwin. 1970. *Microeconomics: Theory and Applications*. New York:

W.W. Norton & Co.

Merrow, Edward W. 1978. *Constraints on the Commercialization of Oil Shale*. R-2293-DOE. Santa Monica, Calif.: The Rand Corporation.

Merrow, Edward W., S. Chapel, and C. Worthing. 1979. *A Review of Cost Estimates in New Technologies: Implications for Energy Process Plants*. R-2481-DOE. Santa Monica, Calif.: The Rand Corporation.

Merrow, Edward W., K. Phillips, and C. Myers. 1981. *Pioneer Plant Study: Analysis of Cost and Performance*. R-2569-DOE. Santa Monica, Calif.: The Rand Corporation.

Merrow, Edward W., and J. Christopher Worthing. 1979. "Possible Shortages in a Synthetic Fuels Mobilization." Unpublished Rand Corporation working paper (prepared under Contract No. DE-AC01-79-7E700078). Cited with special permission.

Monthly Energy Review. May 1981. Monthly energy statistics published by Department of Energy, Energy Information Administration, Washington, D.C.

Mooz, William E. 1978. *Cost Analysis of Light Water Reactor Power Plants*. R-2304-DOE. Santa Monica, Calif.: The Rand Corporation.

———. 1979. *A Second Cost Analysis of Light Water Reactor Power Plants*. R-2504-DOE. Santa Monica, Calif.: The Rand Corporation, December.

Musgrave, R.A., and P.B. Musgrave. 1973. *Public Finance in Theory and Practice*. New York: McGraw-Hill.

National Research Council. 1977. *Assessment of Technology for the Liquefaction of Coal: Summary*. Prepared by the Ad Hoc Panel on Liquefaction of Coal of Committee on Processing and Utilization of Fossil Fuels. Washington, D.C.: National Academy of Sciences.

faction of Coal: Summary. Prepared by the Ad Hoc and Utilization of Fossil Fuels Committee. Washington, D.C.: National Academy of Sciences.

Nordhaus, William D. 1980. "The Energy Crisis and Macroeconomic Policy." *Energy Journal* 1, no. 1 (January): 11-19.

Oak Ridge National Laboratory. 1981. *Liquefaction Technology Assessment— Phase 1: Indirect Liquefaction of Coal to Methanol and Gasoline Using Available Technology*. Study prepared for U.S. Department of Energy. No. ORNL-5665 or U.S. Govt. Printing Office No. 1980-740-062/467. Springfield, VA.: National Technical Information Service.

Office of Technology Assessment. 1980. *An Assessment of Oil Shale Technologies*. U.S. Congress, No. 052-003-0759-2. Washington, D.C.: U.S. Government Printing Office.

OPEC. 1980. "Report of the Group of Experts: Submitted to and Approved by the Fourth Meeting of the Ministerial Committee on Long-Term Strategy." Mimeo of Organization of the Petroleum Exporting Countries (OPEC). Vienna, Austria.

Ostwald, Phillip F., and John R. Reisdorf. 1979. "Measurement of Technology Progress and Capital Cost for Nuclear, Coal-Fired, and Gas-Fired Power Plants Using the Learning Curve." *Engineering and Process Economics* 4,

no. 4, (December): 435-453.
Powell, Stephen G. 1979. "The Effect of Backstop Energy Sources on World Oil Prices." Unpublished paper prepared for Department of Energy Office of Policy and Evaluation.
Ray, Stuart. 1979. "Financial Costing Guidance for the Policy and Fiscal Guidance." Department of Energy memorandum for distribution, March 28. Washington, D.C.
Rowen, Henry, and John Weyant. 1980. "Reducing Our Dependence on Persian Gulf Oil: Needs and Opportunities." Working papers, Stanford University.
Salant, S.W. 1976. "Exhaustible Resources and Industrial Structure: A Nash-Cournot Approach to the World Oil Market." *Journal of Political Economy* 85, No. 5: 1079-1093.
Savay, Albert C. 1976. "Effects of Inflation and Escalation on Plant Costs." *Chemical Engineering* 82 (July): 78-82.
Saymore, Gary. 1980. "The Persian Gulf." In *Energy and Security*, edited by D. Deese and J. Nye, pp. 49-110. Cambridge, Mass.: Ballinger Publishing Company.
Schelling, Thomas C. 1979. *Thinking Through the Energy Problem.* New York: Committee for Economic Development.
Schmalensee, Richard. 1980. "Appropriate Government Policy Toward Commercialization of New Energy Supply Technologies." *The Energy Journal* 1, no. 2 (April): 1-40.
Seidman, David. 1980. "Values in Conflict: Design Considerations for a Two-Stage Synfuels Development Strategy." A Rand Note, N-1469-DOE. Santa Monica, Calif.: The Rand Corporation.
Sobotka & Co., Inc. 1978. "Petroleum Substitutes: Economic Costs and Alternative Subsidy Policies for Shale Oil, Coal Liquids and Methanol." Report submitted under DOE Contract No. EJ-78-C-01-2834. Washington, D.C.
Solow, R.M. 1974. "The Economics of Resources or the Resources of Economics." *American Economic Review.* Papers and Proceedings. Richard T. Ely Lecture.
Stobaugh, R., and D. Yergin, ed. 1979. *Energy Future.* New York: Random House.
Synfuels Interagency Task Force. 1975. *Recommendations for a Synthetic Fuels Commercialization Program.* Volume II: "Cost/Benefit Analysis of Alternate Production Levels," No. 041-001-0111-3. Report submitted to the President's Energy Resources Council. Washington, D.C.: U.S. Government Printing Office.
Tebbetts, Paul. 1978. "Standardized Cost Estimation." Draft Department of Energy report, Office of Policy and Evaluation, December 4. Washington, D.C.
U.S. Senate. 1979. "Extending the Defense Production Act of 1950, as Amended." Report of the Committees on Banking, Housing, and Urban

Affairs and Energy and Natural Resources, to accompany S. 932, Report No. 96-387, October 30. Washington, D.C.

Weinberg, Alvin M. 1979. "Limits to Energy Policy Analysis." Editorial comment in *Energy Policy* 7, no. 4 (December): 274.

Weinstein, M.C., and R.J. Zeckhauser. 1975. "The Optimal Consumption of Depletable Natural Resources." *Quarterly Journal of Economics* (August): 371-392.

White House. 1979. "Fact Sheet on the President's Import Reduction Program." Office of the White House Press Secretary. Mimeo.

INDEX

Actors: and the learning process, 260–262; and incentives, 268–269
Architect, engineering, and construction (AE&C) firms, 8, 111 n.21, 236, 260, 261, 273 n.6, 281–286, 293, 294 n.2

"Backstop" model, 32, 33, 69 n.5/n.6
Badger America, Inc., 273 n.8
Balance of payments, 20
Base case results, 164–175
Base deployment capability (BDC), 52, 53, 139, 188–189
Basic synfuel production cost outcome (SC), 127, 134–135
Bechtel Corporation, 110 n.16, 288, 289, 292, 295 n.8, 296 n.12
Benefit calculation, 149–151
Blum, E.H., 310 n.9
Boomtowns, 71 n.23

Cameron Engineers, 103, 110 n.18, 111 n.21, 194, 282, 294 n.2
Capital charge rate, 77–78, 85–86
Capital intensity, 36
Carter administration, 2
Coal-based synthetic fuels, 10
Coal feedstock costs, 77, 83–84
Coal gasification technology, 88, 267

Coal reserves, 65
Community infrastructure costs, 87
Component suppliers (CS), 260, 261
Constraints, 275–277
Construction labor (CL) services, 260, 286–288
Construction management (CM) firms, 260, 261
Construction phase, 255, 257
Continuing production phase, 256, 258
Conversion efficiency, 84, 108 n.6
Cost-effective program design, 236, 266–267, 272, 275–276, 284–285, 318
Critical mass effect, 5, 52, 53, 71 n.18, 185
Cross-technology learning, 267–268

Decision analysis model, 6–7, 114–117, 303–305; structure of, 119–151
Decision rules, 213–215
Deployability constraints: and factor level inflation, 219 n.9, 229–230
Deployability increase factor (DIF), 54, 139–140, 208, 211
Design phase, 255, 256
Direct production benefits, 191

331

INDEX

Discount rates, 94, 95, 213–215, 222 n.24

Economic benefits, 20
Economic rents, 110 n.14
Economic theory of depletable resources, 32, 109 n.11
"Effective shortfall," 69 n.3
Employment impact, 21, 99
Energy and Security (Hogan), 20
Energy Mobilization Board, 48
Energy Security Act (ESA), 2, 8, 237, 297–299, 308–310
Energy security benefits, 20, 22–28
Enhanced deployability (PBO-D), 5, 7, 47–64, 127, 138–140, 185, 188, 228–229, 275–294
"Expected value," 162, 172–173
"Externalities," 9, 14 n.6, 69 n.8, 212, 215
Exxon Donor Solvent (EDS), 248

Factor cost inflation, 5, 62–63, 74, 98–103, 110 n.19, 144, 153 n.14, 177, 191–203, 229, 320–321; crossover levels of, 195–200, and deployability constraints, 219 n.19, 229–230; and infrastructure development, 200–202, 276, 306–307, 321; and program size, 227–228; and program scale decision, 129–130
"Fast-track" plants, 264–265
Feedstock costs, 77, 83–84
Fluor Corporation, 108 n.6, 289
Ford administration, 2
"Foregone benefits," 230
"Form value," 76, 89; differential for petroleum products, 91–92

Generic technology, 261, 272 n.2
Growth basis, 64–66

Hardware inputs, 277, 288–292
"Hiatus" period, 242–244, 264
Hogan, William, 20, 68 n.1, 71 n.19, 324 n.1
Hyperinflation, 74, 99, 101–103, 200

Inflation, 20, 94, 218 n.1. *See also* Factor cost inflation; Hyperinflation
Information benefits, 4, 5, 17, 28–34, 66–68, 177, 182, 184, 322–323
Information transfers, 249–251, 261–262, 267–268, 271
Infrastructure development, 78, 86–88, 178, 184–185, 275–294, 301; and enhanced deployment capability, 236; and factor cost inflation, 200–202, 276, 306–307, 321; and program size, 228–229; sensitivity analysis of, 207–211
"Interest during construction" costs, 80

Lead-time constraints, 278–279, 292–293
Learning benefits, 5, 7, 38–47, 127, 137–138, 185–188
"Learning curve" concept, 70 n.11, 254, 260, 273 n.5
Learning process: actors in, 260–262
Length of decline period (LDP), 222 n.21
Linkage factor. *See* Deployability increase factor
Lurgi gasification, 247, 248, 272 n.2

Marginal benefits, 222 n.22
Market incentives, 9–10, 36–38, 68, 211–217, 268–269, 308, 315–318
Merrow, E.W., 99–100, 281
Methane production, 88, 247, 267, 272 n.2
Moorehead synthetic fuels bill, 14 n.2

National Constructors Association, 288
"Nominal" synfuel plants, 70 n.14
North Dakota project, 88, 267–268
Nuclear industry, 109 n.8, 273 n.3

Oak Ridge National Laboratory, 108 n.3
Objective function, 213–215
Oil import premium, 3, 56, 60, 68 n.1, 153 n.13
Oil import reductions, 24–25
Oil leverage, 24–25
Oil price jump events (SDE-P), 56–61, 140–142, 177, 183, 203–205, 219 n.8, 227
Oil price (OP) trends, 89–91, 127,

INDEX   333

131–134, 227
Oil pricing benefit, 19–20, 31–34, 66–67, 71 n.22
Oil shale, 10
Operation and maintenance (O&M) costs, 77, 81–83
Organization of Petroleum Exporting Countries (OPEC), 31, 34, 57
Owning and operating company (OOC), 260, 261

Petroleum product costs, 91–92
Phased program structure, 255–256, 264–265, 302–305, 321–322
Plant capacity factor, 78, 86
Plant capital costs, 77, 78–81
Plant facilities investment (PFI), 108 n.2
Portfolio approach, 302
"Positive externalities," 212, 215
Pre-tax discount rate, 213
Price controls, 216
Private investment: and public policy, 11, 28–30, 178, 211–217, 308
Private objective function, 213–215
Procedural constraints, 276–277
Production benefits, 4, 5, 17, 18–28, 66
Production capacity, 70 n.14
Production costs, 6, 21–22, 47, 49–51, 55, 59–60, 76–98
Production date changes, 95–97
Production expansion decision (PED), 144–145
Production goals, 300–301, 302–305, 321–322
Production volume purpose: vs. capability purpose, 237–238
Program benefit outcomes (PBO), 127, 135–140, 177, 184–191
Program scale decision (PSD): and factor cost inflation, 129–131
Program size, 104, 226–229, 302–303
Project Independence, 2
Projects, small, 265–267, 271–272, 324

Rand Corporation, 79, 294 n.2
Rate of time preference, 213
Real resource cost, 76
Resource base: diversity, 247–248; size, 26
Risk, 37, 85, 222 n.23, 269

Sasol project (South Africa), 248, 288
Security events (SDE-S), 56, 61–62, 71 n.20, 127, 142–143, 177, 179–182, 205, 227
Selection learning, 39–41, 67, 137, 239, 240–254, 305–306
Sensitivity analysis, 81, 89–90, 106; cases, 175–218
Service inputs, 277, 281–288
Social benefits and costs, 9, 11, 21–22, 36, 49, 51, 55–56, 64–66, 75, 99, 110 n.14, 178
Social discount rate, 213
Solvent refined coal (SRC), 248
Start-up and initial operation phase, 255–256, 258
Stearns-Roger, 295 n.6
Strategic insurance capability, 4–5, 18, 34–38, 67, 277–281, 300
Sulfur removal technology, 267
Surge deployment benefits, 4–5, 18, 34–68
Surge deployment events (SDE), 127, 140–144, 177, 179–184, 203–207, 227
Synfuel costs: and long-run oil prices, 227
Synfuel production limit (SPL), 127, 145–149, 178, 207–211
Synfuels "startegy" (Hogan), 318–319, 324 n.1
Synthetic fuels (synfuels), defined, 10, 298, 310 n.7
Synthetic Fuels Corporation, 2, 8–9, 237, 297–298, 299–308

"Technological aggressiveness," 269, 306
Technology, new, 248–249
Technology cost distributions, 241–246, 252
Technology diversity, 246, 247
Technology improvement learning, 39, 41–46, 67, 137, 239, 254–269, 306
Technology-specific learning, 261
Timing, 242–244, 264–265, 303–304, 310 n.9

Transportation costs, 87, 88, 109 n.10

U.S. Department of Energy, 85; Policy and Fiscal Guidance (PFG), 89, 132

"Windfall events" tax, 212, 216

ABOUT THE AUTHOR

James K. Harlan is a chemical engineer, economist, and strategic analyst who has been working with synthetic fuels policy and a range of other science and technology policy issues for over five years.

After receiving a B.S. in Chemical Engineering from Washington University in St. Louis, Missouri in 1974, Dr. Harlan turned his professional training toward economics, operations research, and policy analysis at Harvard University's John F. Kennedy School of Government. He received the Master of Public Policy degree in 1977 and completed the PhD in 1981.

Dr. Harlan's professional experience ranges widely including research engineering on automotive emission controls with Exxon, project and program development for a rural and small scale industry technologies at Indonesia's leading technical university (on a Henry Luce fellowship), and assessments of federal cost-sharing policy on synfuels research, development, and demonstration (RD&D) projects for the Office of Management and Budget.

Dr. Harlan's most extensive professional experience has been with the Department of Energy's Office of Policy and Evaluation where from 1977 to 1981 he was responsible for strategic assessments of synthetic fuel programs and industrial energy use issues. He authored several major internal DOE studies including an investment analysis of second generation high Btu coal gasification demonstrations, a coal technology funding strategy, a fossil energy RD&D strategy, and a cost-benefit analysis of initial synthetic fuel production efforts. Following completion of this book as an Adjunct Research Fellow with Harvard's Energy and Environmental Policy Center in September 1981, Dr. Harlan joined the planning staff of the U.S. Synthetic Fuels Corporation where he is the Manager of Strategic Analysis.